2024 NIGHT SKY ALMANAC

A Month-by-Month Guide to North America's Skies from The Royal Astronomical Society of Canada

Nicole Mortillaro

FIREFLY BOOKS

A FIREFLY BOOK

Published by Firefly Books Ltd. 2023
Copyright © 2023 Firefly Books Ltd.
Copyright © 2023 The Royal Astronomical Society of Canada
Text © 2023 Nicole Mortillaro
Photographs © as listed on page 128

First printing

Library of Congress Control Number: 2023935689

Library and Archives Canada Cataloguing in Publication
Title: 2024 night sky almanac : a month-by-month guide to North America's skies from the Royal Astronomical Society of Canada / Nicole Mortillaro.
Other titles: Two thousand twenty-four night sky almanac | Twenty twenty-four night sky almanac | Night sky almanac
Names: Mortillaro, Nicole, 1972- author. | Royal Astronomical Society of Canada, issuing body.
Description: Includes bibliographical references.
Identifiers: Canadiana 2023021617X | ISBN 9780228104322 (softcover)
Subjects: LCSH: Astronomy–Canada–Observers' manuals. | LCSH: Astronomy–Canada–Amateurs' manuals. | LCSH: Astronomy–United States–Observers' manuals. | LCSH: Astronomy–United States–Amateurs' manuals. | LCSH: Astronomy–Popular works. | LCGFT: Handbooks and manuals.
Classification: LCC QB64 .M6722 2023 | DDC 523–dc23

Published in Canada by
Firefly Books Ltd.
50 Staples Avenue, Unit 1
Richmond Hill, Ontario L4B 0A7

Published in the United States by
Firefly Books (U.S.) Inc.
P.O. Box 1338, Ellicott Station
Buffalo, New York 14205

Project manager/Editor (Firefly Books): Julie Takasaki
Project manager (RASC): Robyn Foret, Past President of The RASC
Technical editors: James S. Edgar FRASC (Editor, RASC *Observer's Handbook*) and Chris Vaughan
Graphic design: Noor Majeed
Illustrations and sky charts: Peter Kovalik
Printed in China

Canada

We acknowledge the financial support of the Government of Canada.

Contents

Introduction

If you're reading this, you clearly love the night sky and all the joy it brings you.

This guide aims to provide novice and intermediate amateur astronomers with the knowledge they need to enjoy the night sky in 2024, from understanding planets, nebulae (clouds of gas and dust), comets, asteroids and meteors, to learning about annual and significant celestial events. This is your pocket guide to the cosmos.

But first, it's important that you understand how we navigate the night sky. Just like for navigation on Earth, astronomers use particular coordinates in the sky to figure out what we're looking at.

The **celestial sphere** is an imaginary sphere with Earth at its center. At any one time, an observer's night-sky view only includes half of this sphere, because the other half is below the horizon.

Earth's axis is tilted at 23.5 degrees to the **plane** of the Solar System — the plane being the orbit of the Earth around the Sun. For observers on the ground, the celestial sphere seems to rotate from east to west; that is why the Sun, for example, rises in the east and sets in the west.

The Earth's axis points less than 1 degree away from **Polaris**, the North Star, in the Northern Hemisphere. It's really important to know where north is, so you can navigate around the sky. Use a star chart, your smartphone or both to find true north at your observing location. (In the south, the Earth's axis points about one

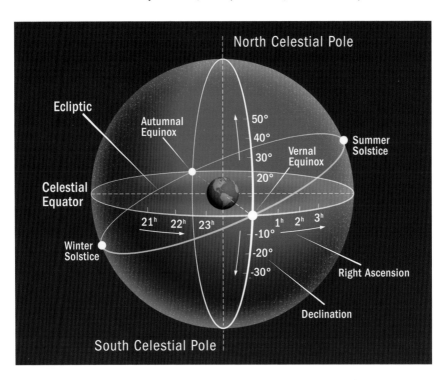

degree away from **Sigma Octantis**, a faint star; given Canada and the United States are in the Northern Hemisphere, however, we will focus on Polaris.) When you look at the sky over long periods of time, Polaris stays still as the other stars rotate around it.

You can easily see this phenomenon by setting up a camera, pointing it toward Polaris, and taking an extended (or long-exposure) photograph of the stars. You will see circular streaks in the sky as other stars rotate around Polaris. Some stars are so close to the North Celestial Pole that they never set. Other stars farther from the pole will rise in the east and set in the west, just like our Sun. Many stars are so far from the pole that they never rise above your horizon.

While beginner astronomers will navigate using a simplified star chart, it's helpful to know some of the terminology we use for finding our way around the sky. That way, as you gain confidence, you can follow professional astronomers in their observations.

The point directly overhead your observing location is called the **zenith**, and the **celestial equator** is directly above the Earth's equator. The point directly below you (under the horizon and opposite to the zenith) is the **nadir**. The line that runs from the north point on the horizon, through the zenith, to the south point on the horizon, is the **meridian**.

To navigate the night sky, astronomers created a type of latitude and longitude called **celestial coordinates**. In astronomy, we use **declination** (Dec.) and **right ascension** (RA). Declination is like latitude on Earth, running from north to south. Right ascension is like longitude on Earth, running from west to east. Declination is measured in degrees, and right ascension is measured in hours, minutes and seconds.

Star trails with Polaris at the center

As Earth rotates, the apparent position of most of the stars changes. More advanced astronomers may want to learn more about exactly where to find faint stars in their telescopes. For these situations, we navigate using **altitude** (angular elevation above the horizon, between 0 degrees at the horizon and 90 degrees at the zenith) and **azimuth** (the number of degrees clockwise from due north). But if you're just starting out, don't worry yet about these advanced observing techniques. A simple star chart will help you find the constellations and the planets.

If you want to look for planets, you should also pay attention to the **ecliptic**. That is the path the Sun takes through the constellations, and the Moon and planets do not stray far from that path.

Handy Sky Measures

Astronomers need to know how far apart things are in the sky. And they do this using angular degrees.

It's not hard to imagine the sky as a sphere that measures 360 degrees — after all, space surrounds Earth on all sides. Standing in one spot on Earth, if you trace the sky from horizon to horizon, that would equal 180 degrees. Remember, the other 180 degrees is under the horizon.

If you want to measure the distances between two objects — say, between the Moon and Venus as they appear together in the sky — you can use your hand as a measuring tool. It all lies within your fingers.

Hold your hand at arm's length. The width of your pinky finger equals 1 degree. The width of your three middle fingers held at arm's length equals five degrees; a closed fist is 10 degrees; the distance between the tip of your index finger and the tip of your pinky is 15 degrees; and the distance between your thumb and pinky is roughly 25 degrees.

This is particularly helpful when trying to see how high or low something is above the horizon.

You can practice measuring the degrees with stars found in the Big Dipper. The chart shows how to find the Big Dipper using Polaris. Because it's circumpolar, the orientation of the Big Dipper varies.

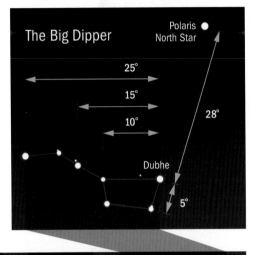

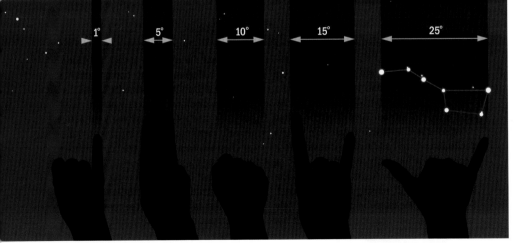

A view of Polaris and the Big Dipper
over Dinosaur Provincial Park, Alberta

The Night Sky: A Cosmic Time Machine

Every time you look up at the sky, you are looking back in time.

Light from the Sun takes 8.3 minutes to reach us. Light from the Moon takes roughly 1.3 seconds. That's because light takes time to travel. In fact, light travels through space at roughly 300,000 kilometers (186,000 miles) per second.

We call the distance light travels in one year a **light-year**. So, when we look at stars, galaxies and nebulae, they appear how they looked back in time, depending on their distance to us. For example, the Andromeda Galaxy is the closest spiral galaxy to our own Milky Way. It is easily visible to the unaided eye in the Northern Hemisphere at dark-sky sites, even though it lies roughly 2.5 million light-years from us. That means, we are seeing the galaxy as it was 2.5 million years ago.

Sirius, also known as the "Dog Star," is the night sky's brightest star. It's visible during the winter months in the Northern Hemisphere. Its distance of 8.6 light-years from Earth means we are looking at it as it was 8.6 years ago.

Our Universe is roughly 13.8 billion years old. Astronomers improve their understanding of the evolution of our Universe by using powerful telescopes that can see farther and farther away, and so farther back in time.

Launched in 1990, the Hubble Space Telescope has provided us not only with jaw-dropping images of astronomical objects such as

Andromeda Galaxy (Messier 31), our closest galactic neighbor

Captured by NASA's James Webb Space Telescope and titled "Webb's First Deep Field," at the time this image was the deepest and sharpest infrared image of the distant Universe

galaxies and nebulae, but it has also helped narrow down the age of the Universe.

And more recently, in 2021, the James Webb Space Telescope opened its eyes for the first time, providing us with a look back to some of the very first galaxies to form — back to when the Universe was less than a billion years old.

It's almost like looking at a baby picture of our Universe.

While time machines may not be possible, every time we pick up a pair of binoculars or peer through our telescopes, we are taking a voyage back in time.

Binoculars and Telescopes

Using your eyes to move around the sky allows you to see the Moon, the planets, the Sun and even some objects outside of our Solar System. However, having binoculars or a telescope opens up the Universe to you.

Binoculars are an amazing first tool for seeking out the marvelous wonders of the night sky. A good pair of binoculars can reveal the intricacies of the Moon, showing its peaks and valleys, or uncover the daily motion of Jupiter's moons as they dance around our Solar System's largest planet. A decent telescope can make these objects appear bigger and show details of star clusters or nebulae.

But the question that often arises is what type of binoculars or telescope should I get?

For binoculars, it's important to understand two things: the **magnification** (or power) and the **aperture**. Binoculars are usually represented by two numbers, separated by an "×." Two example binocular numbers are 7×50 or 10×50. The magnification is the first number, and it represents the number of times larger something will appear compared to viewing it with the naked eye. The second number is the aperture in millimeters (about 3/64ths of an inch), and it represents the diameter of each lens. Good viewing in part depends on how well your binoculars can gather light. If your binoculars have a larger second number, they have a larger aperture and can gather more light from distant objects. But that comes with a cost. Bigger binoculars are heavier and, therefore, more difficult to hold, which means you will need a tripod to steady your view.

Another number often given for binoculars is the **field of view**, or FOV. This is how wide you will be able to see in degrees (see "Handy Sky Measures," on page 6). In general, the higher the magnification you use, the smaller your field of view will be. At times, this will mean you need to make a choice. Do you want to zoom in on a particular crater on the Moon, for example, or observe a more spread-out chain of mountains? Making these decisions is difficult for astronomers, too, so don't worry if it feels hard — it gets

A pair of binoculars is a great portable tool for observing the night sky

a bit easier to decide what to do with practice.

The top end of handheld binoculars is typically 7×50. Larger than that, it's best to get a tall tripod if you want the best view possible and to share views with others.

We recommend using binoculars for a few months before investing in a telescope. Once you are comfortable with binoculars, luckily, your knowledge will be useful because telescopes use many of the same definitions for observing.

Remember that light-gathering ability is important; a typical beginner's telescope usually ranges between 50 to 150 millimeters (2 to 6 inches) in **aperture**, the diameter of the main lens or mirror. The bigger the aperture, the more light the lens will gather, and the brighter the view will be. Invest in a sturdy tripod, too, so that the telescope will not shake when you use it.

The **magnification** of a telescope depends on the eyepiece used; magnification is calculated by dividing the focal length of the telescope by the focal length of the eyepiece (using like units). For example, a telescope with a focal length of 1,000 millimeters and a 10-millimeter eyepiece will magnify an object 100 times. The same 10-millimeter eyepiece in a 500-millimeter telescope would magnify an object 50 times.

While a bigger aperture and higher magnification may seem like the best way to go, it's best not to get too caught up in choosing a telescope with the highest numbers. What is important is how you plan to use it. Since most people are unlikely to have backyard observatories, the most important thing may be portability. Too heavy a telescope means you're unlikely to haul it out to the backyard or to a vacation spot, far from city lights.

When it comes to telescopes, there is a wide array of choices. Some of the most popular

A refractor telescope, which is best used for viewing planets, the Moon and double stars

are refractors, reflectors and compound telescopes. Each type has its own benefits, and it all depends on what you prefer. We therefore recommend you try testing out different telescopes at a local star party held by astronomy groups before making the investment. Alternatively, check reputable astronomy magazines or forums for recommendations for beginners; in most cases, all it takes is a little Internet searching and some patience.

A **refractor telescope** has a front lens that focuses light to form an image at the back, and an eyepiece that acts like a magnifying glass to allow your eye to focus on that image. Refractors tend to be more reliable, as their lenses are fixed in place and, therefore, don't get out of alignment as easily as some other types of telescopes. Refractors are best used for planetary and lunar observing as well as viewing double stars.

A **reflector telescope** uses a mirror to gather and focus light. The advantage for reflectors is they don't usually suffer from **chromatic aberrations**, when light of different wavelengths (i.e., colors) doesn't focus on the same point; this makes them ideal to observe distant objects like star clusters. Chromatic aberrations cause the different colors of the spectrum to split and the image to appear blurry, which isn't great when looking at a group of stars. A **Dobsonian telescope**, which is a variant of a reflector with a simple mount, gives you a far bigger aperture for far less than the cost of other telescopes.

Compound or **catadioptric telescopes** combine lenses and mirrors to form an image. The **Schmidt-Cassegrain**, a type of compound telescope, has a compact design that makes it quite popular. With this instrument, an astronomer can get a bigger aperture in a smaller sized telescope. This telescope type is also portable, making it easier to move the equipment into remote areas.

Star parties are great opportunities to check out different telescopes and chat with fellow enthusiasts

A reflector telescope (front) and a variant of a reflector telescope, a Dobsonian telescope (back), are ideal for viewing distant objects since they don't usually suffer from chromatic aberrations

A Schmidt-Cassegrain telescope, which uses both lenses and mirrors, is compact and best for viewing finer details on planets and the Moon

Stars

Stars come in many different varieties. Our Sun is a **yellow dwarf star**, on the **main sequence**, meaning that it's converting hydrogen into helium at its core, like most other stars. When a star does this conversion, it releases a tremendous amount of energy. This energy, in the form of sunlight, allows life to thrive on Earth.

The Sun, at 4.5 billion years old, is considered middle-aged. When it runs out of fusion fuel, five billion years from now, it will at first swell, becoming a **red giant** and engulfing the inner planets, then it will slough off its outer layers to become a **white dwarf**.

One of the most interesting types of stars, some might argue, are **red supergiants**. These colossal stars — roughly 1,400 times the mass of the Sun — have relatively short lifespans.

And when supergiants do stop fusion, they do so in a spectacular fashion, in an explosion called a **supernova**. A supernova occurs when a star can no longer convert hydrogen into helium; eventually the core is converted into iron. During a supernova, the star first collapses and then explodes outward, creating even heavier elements.

Betelgeuse, a star found in the left shoulder of Orion, is a red supergiant. While many far-away and faint supernovae have been witnessed in modern history in other galaxies, none have occurred in our own galaxy — the **Milky Way**. When Betelgeuse goes supernova, its brightness will rival the full Moon in our sky. If you're hoping to see Betelgeuse explode, you're not alone. However, estimates peg Betelgeuse's death for some time within the next 100,000 years.

The most common star in the Universe is a **red dwarf**, which is a cool star that's much smaller than the Sun. Astronomers often search for potentially habitable planets around these stars, because the inherent dimness and smaller size of red dwarfs makes it easier to spot planets. The stars, however, can be quite volatile, occasionally releasing a tremendous amount of radiation. Intense radiation is not a friendly process for most life forms.

The **Hertzsprung-Russell diagram** (H-R diagram) was developed in the early 1900s by Ejnar Hertzsprung and Henry Norris Russell, following the research findings by two Harvard computers, Annie Jump Cannon and Antonia Maury. The diagram plots the temperature of stars against their luminosity. Just as we do, stars go through certain stages in their lives.

Top 10 Brightest Stars in the Night Sky	
1.	Sirius
2.	Canopus
3.	Alpha Centauri A (Rigil Kentaurus)
4.	Arcturus
5.	Vega
6.	Capella
7.	Rigel
8.	Procyon
9.	Achernar
10.	Betelgeuse

The Harvard Observatory Computers

In 1881, Edward Charles Pickering, the director of the Harvard Observatory, hired a team of women to compute and catalog photographs of the night sky. Their work was incredibly important in providing the foundations of astronomical theory. Annie Jump Cannon, one of the computers, added to work done by fellow computer Antonia Maury and developed a system of classifying stars that is still used today.

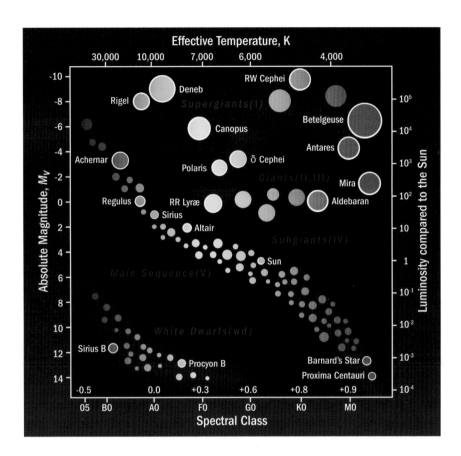

The H-R diagram provides astronomers with the information about a star's current age. Main-sequence stars that are fusing hydrogen into helium — such as our Sun — lie on the diagonal branch of the diagram.

Finally, there's a star's **magnitude**, or apparent brightness. Magnitude is measured on a scale where the higher the number, the fainter it is — and negative numbers are brighter than positive numbers. For example, Sirius, the brightest star in the night sky, measures -1.4 on this scale. Polaris is 2.0, and the Sun is -27. We also use magnitude to measure the brightness of other celestial objects, like the

Moon, planets, asteroids and comets.

Note there is a difference between apparent and absolute magnitude. Apparent magnitude is the brightness of an object that we observe from Earth, but absolute magnitude is the brightness of an object if it were placed 32.6 light-years from Earth. This second measure helps astronomers directly compare the luminosity of objects and is what is used for the H-R diagram. On this scale, Sirius has a magnitude of 1.4, Polaris is -3.6 and the Sun is 4.8.

In astronomy, Greek letters of the alphabet are used to identify stars within a constellation usually from brighter to dimmer.

Constellations

Constellations, groups of stars that make imaginary images in the night sky, have been around since ancient times. Typically, the images astronomers use are based on Greek, Roman and Arabic mythologies. The International Astronomical Union (IAU) recognizes a total of 88 official constellations.

Some of the most recognizable constellations in the Northern Hemisphere are Orion (the Hunter), Cygnus (the Swan), Leo (the Lion), Gemini (the Twins), Scorpius (the Scorpion) and Ursa Major (the Great Bear).

More recently, there has been more effort to acknowledge constellations of Indigenous peoples. The naming of Indigenous constellations varies around the world, making these constellations regionally distinct. Some groups see Ursa Major as a bear, or a caribou. The Cree see Corona Borealis as seven birds, and Cepheus as a turtle. To the Navajo, Polaris is Nahookos Bikq, meaning central fire. For some Indigenous peoples, such as the Inuit, the Northern Lights are dancing spirits.

Aside from the constellations, there are also **asterisms**, a group of stars within a constellation (or sometimes from several different constellations) that forms its own distinct pattern. The Big Dipper is probably the most famous asterism, as its stars lie within the constellation of Ursa Major. There's also the Summer Triangle, with bright stars from Cygnus, Lyra (the Lyre) and Aquila (the Eagle), and the Winter Triangle with stars from Orion, Canis Major (the Great Dog) and Canis Minor (the Little Dog).

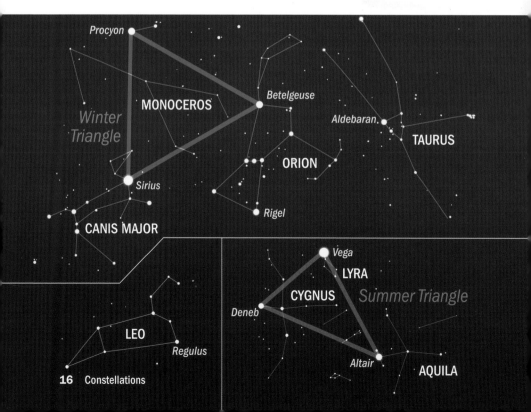

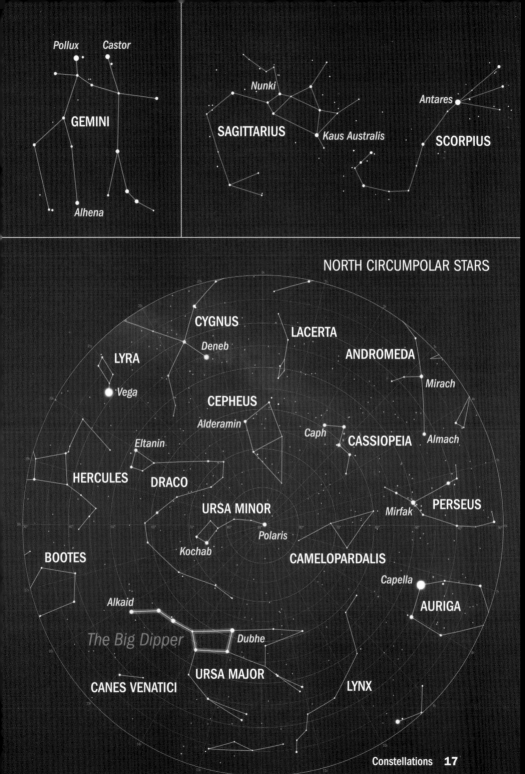

GEMINI

Pollux Castor

Alhena

SAGITTARIUS

Nunki

Kaus Australis

SCORPIUS

Antares

NORTH CIRCUMPOLAR STARS

CYGNUS

Deneb

LACERTA

ANDROMEDA

LYRA

Vega

Mirach

CEPHEUS

Alderamin

Caph

CASSIOPEIA

Almach

Eltanin

HERCULES

DRACO

URSA MINOR

Mirfak

PERSEUS

Polaris

Kochab

CAMELOPARDALIS

BOOTES

Capella

AURIGA

Alkaid

The Big Dipper

Dubhe

CANES VENATICI

URSA MAJOR

LYNX

Comets, Asteroids and Meteors

Comets are icy balls of debris — specifically, dust and ice — left over from the formation of our Solar System. They are sometimes referred to as "dirty snowballs" and can be stunning objects to see in the night sky when tails of dust and ionized gas fan out behind their cores. While we know of many comets, predicting the appearance of a bright one is all but impossible. Comets tend to fall apart before becoming too bright. The Sun's gravity and heat are strong, and comets themselves are very delicate visitors from the outer Solar System or beyond. Many comets literally crumble under the pressure as they dive in toward the Sun and the inner Solar System, where our planet resides.

Bright comets can be marvelous. On August 17, 2014, Comet Lovejoy C/2014 Q2 was spotted as it came toward the inner Solar System. This comet produced a spectacular tail as it neared the Sun. Tails occur when ice **sublimates**, turning directly from a solid into a gas.

There have been other wonderful naked-eye comets that have graced our night sky, including Comet Hyakutake C/1996 B2, Hale-Bopp C/1995 O1, and Comet NEOWISE C/2020 F3. Each comet has the year of its discovery in its official name, so NEOWISE was found in 2020, Hyakutake in 1996, and so on. Many Northern Hemisphere observers were treated to a special show when Comet NEOWISE appeared in the skies in 2020. It was one of the brightest comets to appear in a generation — even city dwellers could spot it through light pollution.

Periodic comets, or ones that we can predict, are given the designation "P," while those that appear unexpectedly are given the designation "C." For example, Halley's Comet, or 1P/Halley, appears roughly every 76 years — a clear P. The last time it passed was 1986; the next time will be in 2061. In general, C comets are brighter than P comets because P comets have sublimated their material into space from repeated trips by the Sun.

Like comets, **asteroids** (which are called minor planets) are left over from the formation of our Solar System. Instead of dust and ice,

Comet NEOWISE C/2020 F3

asteroids are rocks and come in all shapes and sizes. Most travel in the asteroid belt between Mars and Jupiter; however, they can be found other places, too, including beyond the orbit of dwarf planet Pluto at the outer edge of the Solar System, in a region called the Kuiper Belt. There are three broad composition types of asteroids:

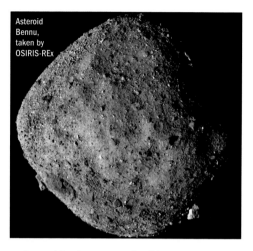

Asteroid Bennu, taken by OSIRIS-REx

- **Chondrite (C-type)**, which are made up of clay and silicate rocks. They are dark in appearance and the most common. They are also the oldest type of asteroid.
- **Nickel-iron (M-type)**, which are believed to have experienced high temperatures and melted after they formed.
- **Stony (S-type)**, which are made up of silicates and nickel-iron.

Space agencies have sent spacecraft to collect samples from a few asteroids and return them to Earth. These samples can help us better understand the formation of our early Solar System. NASA's OSIRIS-REx spacecraft touched down on an asteroid called Bennu in October 2020 and left in May 2021 to deliver samples of Bennu to Earth. The samples arrived in September 2023 and soon after the spacecraft began a new mission as OSIRIS-APophis EXplorer (OSIRIS-APEX) and headed toward an asteroid called Apophis.

Like comets, asteroids can sometimes be knocked out of their orbits. In the asteroid belt, their orbit may be influenced by Jupiter's massive gravity. In the Kuiper Belt, they may be disturbed by interacting with other asteroids.

Asteroids known as centaurs travel in between the orbits of Jupiter and Neptune. Some of them may eventually be ejected from the Solar System, or travel past the inner planets (Mercury, Venus, Mars and Earth), or even burn up if they get too close to the Sun.

Asteroids that cross Earth's orbit are known as **potentially hazardous asteroids (PHAs)** or **near-Earth objects (NEOs)**. Asteroid impacts are a serious business. Sixty-six million years ago, an asteroid roughly 10 kilometers (6 miles) wide slammed into the area that is now the Yucatan Peninsula of Mexico. This event, sometimes referred to as the Cretaceous–Paleogene extinction, caused 75 percent of all animals on Earth to die out, including the dinosaurs.

While our planet does still exist in a shooting gallery (there are billions of objects in the Solar System, ranging from dust grains to kilometer-sized rocks), there are many organizations — including NASA and the European Space Agency — that are constantly searching for any objects that could cross Earth's orbit. To date, NASA says that it has identified nearly 90 percent of any objects 1 kilometer (0.6 miles) wide or larger. And the good news is that we currently know of no PHAs that are a threat to life on Earth up until the year 2200.

But that doesn't mean scientists aren't trying to prepare for the unexpected.

In the early morning hours of February 15, 2013, the sky over Chelyabinsk, Russia, became awash in light. A previously undetected object roughly 20 meters (66 feet) wide entered Earth's atmosphere and exploded in the sky, producing an airburst that shattered windows and injured more than 1,000 people.

Scientists are looking for ways to better detect these somewhat smaller objects, which may not be planet-destroying but could cause harm to populated areas. And they're even looking for ways to change their orbits.

In 2021, NASA conducted the first experiment trying to do just that. It launched its Double Asteroid Redirection Test, or DART. The small spacecraft traveled to a **binary asteroid** — a small asteroid called Dimorphos orbiting a larger asteroid called Didymos. The object of the test was to smash a spacecraft into Dimorphos to see if the impact could change its orbit. If humans can use a spacecraft to alter a PHA's orbit, it will no longer pose a threat to Earth.

The mission was a success. Before the impact, it took Dimorphos 11 hours and 55 minutes to orbit Didymos; after the impact, it took 11 hours and 23 minutes. Photographs by Hubble even showed that Dimorphos formed twin tails for some time after the impact (see the photo on the opposite page).

A **meteoroid** (from the Greek *meteoros*, meaning "high in the air") is a small piece of debris that has broken off from a larger object, usually an asteroid or a comet. Meteoroids are usually the size of a small pebble, but they can be smaller or a little larger. A **meteor** is the light, heat and (occasionally) sound phenomena produced when a meteoroid, collides with molecules in Earth's upper atmosphere. We also call meteors **shooting stars**.

When a meteoroid enters Earth's atmosphere, the surface of the object is heated. Then, at a height typically between 119.8 and 79.9 kilometers (74.5 miles and 49.7 miles), the meteoroid begins to **ablate**, or lose mass. Meteoroid ablation usually happens through vaporization, although some melting and

The vapor trail left after an object entered Earth's atmosphere over Chelyabinsk on February 15, 2013

breaking apart can also occur. If a meteoroid is large enough that it doesn't burn up entirely and reaches the ground, it is called a **meteorite**.

Meteoroids can be divided into two groups: **stream** and **sporadic meteoroids**. Stream meteoroids have orbits around the Sun that often can be linked to a parent object — in most cases a comet, but it can sometimes be an asteroid. When Earth travels within the orbit of a stream, we get a **meteor shower**. Because all the objects (meteoroids) move in the same direction, they all seem to come from one point in the sky, called the **radiant**. The constellation containing the radiant (or in some cases a nearby star) is what gives a meteor its name, such as the Perseids or Geminids (two of the most active meteor showers).

Sporadic meteors occur in random parts of the sky, with no radiant.

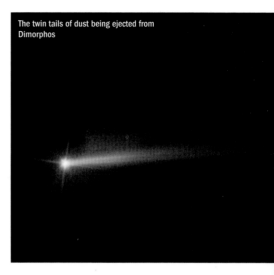

The twin tails of dust being ejected from Dimorphos

The Geminid meteor shower

Meteor Showers in 2024

There is nothing more wonderful than stepping out and looking up at the night sky, only to see a brief streak of light flash against the stars.

Almost monthly, we get major meteor showers. Some showers are best seen from the Northern Hemisphere, while some are better visible in the south, depending on the radiant.

When astronomers talk about the "peak" of a meteor shower, they're referring to the **Zenithal Hourly Rate**, or the ZHR. This is the rate of meteors a shower would produce per hour under clear, dark skies and with the radiant at the zenith. In practice, a single observer will likely see considerably fewer meteors than the ZHR suggests, owing to the presence of moonlight, light pollution, the location of the radiant, poor night vision and inattentiveness.

Here is a list of major meteor showers that you can enjoy simply by staying warm and looking up. No special equipment is needed beyond your eyes; just make sure to give yourself about 20 minutes to adjust to the darkness before searching for meteors.

Quadrantids: December 27, 2023–January 10, 2024

The Quadrantids might be one of the best meteor showers of the year, with a ZHR of 120 for a brief interval of time. The only thing holding it back from earning the title is that January tends to be cloudy over North America, and the shower's peak has a brief window of six hours, making the average hourly rate closer to 25. But there's good news: the meteors often produce bright **fireballs**, or bright meteors with a magnitude higher than −4.0. This shower's radiant lies between Boötes and Draco. The Moon will be just after last quarter on the peak night.

Parent Object: Asteroid 2003 EH1
2024 Peak Night: January 3–4

Lyrids: April 16–30, 2024

After a dearth of meteor showers in the months of February and March, April brings us the Lyrids. This shower produces a ZHR of 18, so it's not a particularly strong shower, but the meteors that do appear tend to produce fireballs. A nearly full Moon will interfere with all but the brightest meteors.

Parent Object: Comet C/1861 G1 (Thatcher)
2024 Peak Night: April 21–22

Eta Aquariids: April 19–May 28, 2024

The Eta Aquariids aren't a stellar show for the Northern Hemisphere, since the radiant rises in early dawn, but they can still produce a ZHR of 20 meteors, if you care to get up early enough. Rather than fireballs, this shower tends to produce long trains through the sky. The Moon won't interfere with observations on the peak night, as it will be a waxing crescent.

Parent Object: Comet 1P/Halley
2024 Peak Night: May 4–5

Perseids: July 17–August 26, 2024

For those in the Northern Hemisphere, the Perseids are considered the best show of the year. The weather is warm and there are fewer clouds at this time of year. The shower's ZHR is 100, though rates of 50 to 75 are more commonly seen on the peak night. The great news is that the Moon will not interfere with the show.

Parent Object: Comet 109P/Swift-Tuttle
2024 Peak Night: August 11–12

Orionids: October 2–November 7, 2024

This shower is considered medium strength, though it can sometimes surprise us with more activity. On average, the Orionids produce a ZHR of 10 to 20 meteors, though from 2006 to 2009, the shower produced roughly 50 to 75 an hour. The meteors that enter our atmosphere are fast — roughly 66 kilometers (41 miles) per second — and, as a result, produce long, glowing trains. Fireballs are also possible. The bad news is that the Moon will be roughly 80 percent illuminated.

Parent Object: Comet 1P/Halley
2024 Peak Night: October 20–21

Southern Taurids:
September 10–November 20, 2024
The Southern Taurids last for two months and have several minor peaks. Though this shower produces a ZHR of only 5, it can produce some fireballs. The shower is stronger in the Southern Hemisphere, though we can also catch a few meteors in the north. The Moon will not interfere with this shower on the peak night.

Parent Object: Comet 2P/Encke
2024 Peak Night: November 4–5

Northern Taurids:
October 20–December 10, 2024
Like the Southern Taurids, the Northern Taurids last roughly two months and have a ZHR of 5. There are often reports of an increase of fireballs during the period when the two showers overlap. A crescent Moon will slightly interfere with this shower.

Parent Object: Comet 2P/Encke
2024 Peak Night: November 11–12

Leonids: November 6–30, 2024
While the Leonids don't produce a high number of meteors an hour (they have a ZHR of 15), they can produce **outbursts**, or meteor storms; the most recent such outbursts occurred in 1999 and 2001. Unfortunately, the next outburst isn't expected until 2099. Still, the Leonids do put on a show, with bright meteors and long-lasting trains. An almost-full Moon will wash out all but the brightest meteors.

Parent Object: Comet 55P/Tempel-Tuttle
2024 Peak Night: November 16–17

Geminids: December 4–17, 2024
Due to the weather — with increasing cloudiness and chillier temperatures — December isn't an ideal time to catch a meteor shower. But December is the month with the most active meteor shower of the year, called the Geminids. This shower has a ZHR of 150, and the meteors it produces are usually bright and colorful. The downside is that the Moon will be almost full.

Parent Object: Asteroid 3200 Phaethon
2024 Peak Night: December 13–14

Ursids: December 17–26, 2024
Right on the heels of the Geminids is the often-forgotten Ursid meteor shower. The ZHR for this shower is 10, but sometimes you might catch an outburst that could produce a ZHR of 25. The Moon will not interfere with this shower.

Parent Object: Comet 8P/Tuttle
2024 Peak Night: December 21–22

The Moon

The Moon is most often the first astronomical object to capture anyone's attention in the night sky. It's also likely the first object most people see up close, whether it be through binoculars or a telescope.

Our nearest celestial neighbor has been an object of fascination since humans first looked up to the sky. Early scientists in ancient observatories carefully tracked the cycles of the Moon. Some 23 centuries ago, Aristarchus of Samos carefully observed a total lunar eclipse and derived impressively accurate measurements of the Moon's diameter. He estimated its diameter was one-third that of Earth. He was close: the actual percentage is 27.2, and its diameter is roughly 3,540 kilometers (2,200 miles).

It takes about 29.5 days for the Moon to go through its cycle of **phases**. When you can't see the Moon at all, it's called a **new Moon**.

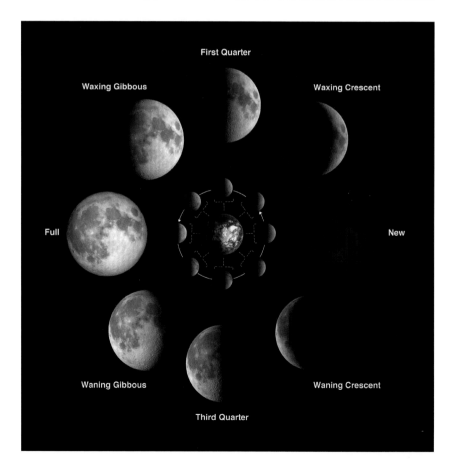

First Quarter

Waxing Gibbous

Waxing Crescent

Full

New

Waning Gibbous

Waning Crescent

Third Quarter

When the Moon is fully illuminated, it is a **full Moon**. The list of phases is new Moon, waxing crescent Moon (when the Moon is just a thin crescent), first quarter Moon (when the eastern half is lit), waxing gibbous Moon (when the Moon is between half-full and full), and then full Moon. The process runs in reverse from full to new: waning gibbous Moon, third or last quarter Moon (lit on the western half), waning crescent Moon, and new Moon.

The Moon is tidally locked with Earth, meaning that we only ever get to see one face of it. The Sun does shine on the other side of the Moon when we can't see it, so don't call the far side of the Moon the dark side.

While it may seem that we always see the same Moon features month after month, that's not entirely true. The Moon oscillates from our perspective, a process called **lunar libration**. As a result, at certain times we see small zones along the Moon's edge that are normally hidden.

Observing the Moon

You don't need a telescope to enjoy the Moon. If observed even briefly on a regular basis, Earth's only natural satellite has much to offer the naked-eye observer. Keep an eye on it, and you will notice things like its wandering path through the constellations, the changing phases, frequent **conjunctions** (where two objects appear close together in the sky) with planets or bright stars, occasional eclipses, lunar libration, earthshine (illumination on the Moon reflected from our planet) and other wonderful atmospheric effects.

If you happen to take a look through a telescope, binoculars or even a camera, our closest astronomical target offers an amazing amount of interesting detail. There are countless features on the lunar nearside, and more than 1,000 have been formally named by the IAU. Many of the greatest names in astronomy, exploration and discovery are commemorated through these names. For extra dramatic effect, try looking at a lunar feature when the **terminator** — the line between darkness and sunlight — runs nearby the feature. The extra shadow could make it easier to see details in the object you're interested in looking at.

An estimated three to four billion years ago, our Solar System was bombarded by debris for hundreds of millions of years. This period is called the Late Heavy Bombardment, and you can see many of the scars from that time left over on the Moon.

Because the Moon has almost no atmosphere and no tectonic activity, those scars have remained over billions of years. Some of that activity created deep **impact basins** that filled with lava flows and hardened into basaltic rock. These largely circular regions were termed **maria**, or seas, by early astronomers because they resembled Earth's oceans.

The brighter white areas, which are highlands that surround the maria and dominate the southern portion of the Earth-facing hemisphere, feature many ancient **impact craters**.

After the bombardment slowed to a trickle, the Moon remained relatively free from intense impact activity — meaning the Moon has changed very little since billions of years ago. But, just like on Earth, meteorites still periodically slam into the Moon today.

The Moon map on the next two pages shows a few interesting features for you to target as you observe the Moon.

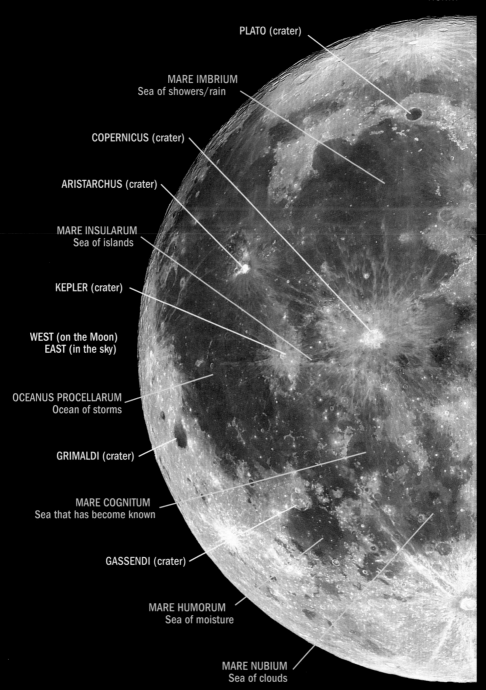

NORTH

PLATO (crater)

MARE IMBRIUM
Sea of showers/rain

COPERNICUS (crater)

ARISTARCHUS (crater)

MARE INSULARUM
Sea of islands

KEPLER (crater)

WEST (on the Moon)
EAST (in the sky)

OCEANUS PROCELLARUM
Ocean of storms

GRIMALDI (crater)

MARE COGNITUM
Sea that has become known

GASSENDI (crater)

MARE HUMORUM
Sea of moisture

MARE NUBIUM
Sea of clouds

SOUTH

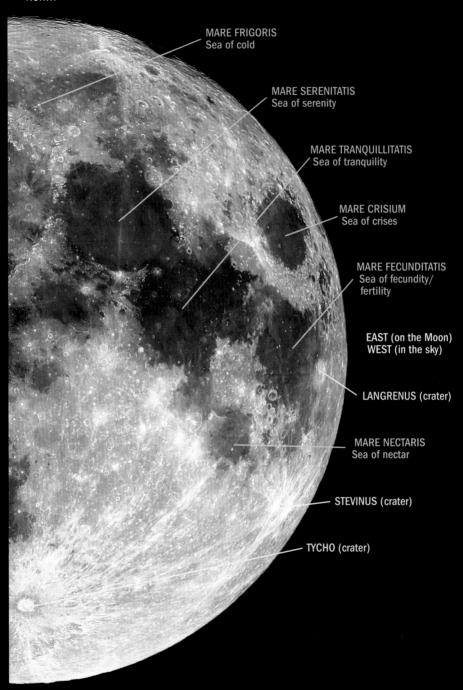

NORTH

MARE FRIGORIS
Sea of cold

MARE SERENITATIS
Sea of serenity

MARE TRANQUILLITATIS
Sea of tranquility

MARE CRISIUM
Sea of crises

MARE FECUNDITATIS
Sea of fecundity/
fertility

EAST (on the Moon)
WEST (in the sky)

LANGRENUS (crater)

MARE NECTARIS
Sea of nectar

STEVINUS (crater)

TYCHO (crater)

SOUTH

The Sun

Our Sun highly influences Earth. It is connected with our seasons, weather, ocean currents and climate, and its power makes life itself possible.

Unlike Earth, the Sun isn't a solid body, and due to that, different parts of it rotate at different rates. At the equator, for example, it spins roughly once every 25 days, while the polar regions take about 30 days to complete a rotation.

In the core of the Sun, where hydrogen atoms fuse to make helium, temperatures are a searing 15 million degrees Celsius (27 million degrees Fahrenheit). The tremendous amount of energy generated at the core — **thermonuclear fusion** — is carried outward by radiation, taking roughly

Take Care of Your Eyesight

Do not observe the Sun without special protective equipment, such as a certified solar filter that covers your eyes or telescope aperture entirely. Unprotected observations even for a few seconds could damage your eyesight permanently.

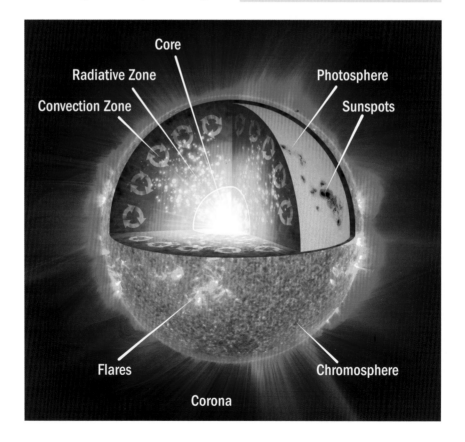

Core
Radiative Zone
Convection Zone
Photosphere
Sunspots
Flares
Chromosphere
Corona

170,000 years to get from the core to the top of the convective/convection zone. As hot as the core is, at the surface, it's a much different story — the temperature is a much "cooler" 5,500 degrees Celsius (10,000 degrees Fahrenheit).

The Sun also has a roughly 11-year cycle, that has a **maximum** (a period of increased solar activity) and a **minimum** (a period of low solar activity). During a maximum, the number of **sunspots**, which are cooler regions on the surface of the Sun, increases. Sometimes, the magnetic field lines of these sunspots can become entangled, finally snapping and releasing a tremendous amount of radiation into space. This process is called a **solar flare**. And often a **coronal mass ejection (CME)** can follow a flare, with charged particles of the Sun speeding outward into space. If these particles reach Earth, they can disrupt radio transmissions, damage satellites and, more positively,

A New Solar Cycle

At the end of 2020, we started Solar Cycle 25. After a quiet solar minimum, we can now expect to see more solar activity. Solar Cycle 24 was the weakest cycle in the last 200 years.

interact with our magnetic field. When the particles do interact with the field, they can produce an aurora.

Though beautiful, it must be said that these outbursts from the Sun can also cause power outages, as was experienced in Quebec in 1989. As such, astronomers are keen to better understand our nearest star using many spacecraft — such as the Parker Solar Probe, which was launched in 2018. Keep watching the eight-year adventure of the Parker Solar Probe as it studies the Sun's activity.

The Northern Lights

Observing the Sun

The Sun, our closest star, is a wonderful object to observe with any type of telescope, if used safely. The only safe filter is the kind that covers the full aperture of the telescope, at the front, allowing only 1/100,000th of the sunlight through the telescope. Without this filter, you risk permanently damaging your telescope or, far worse, your eyes. Ensure that the filter material is specifically certified as suitable for solar observing, and *take great caution every time you observe the Sun.*

The preferred solar filters are made of Thousand Oaks glass or Baader film. Thousand Oaks glass gives the Sun a golden-orange color, while Baader film gives the Sun a more natural white look, which can provide good contrast if there are any bright spots, or **plages**, on the Sun. Other solar-filter options include lightweight Mylar film (which gives the Sun a blue-white tint), and eyepiece projection. Projecting the image from the eyepiece onto white cardstock paper, or a wall, produces an image that can be safely shared with others.

While it might seem like the Sun is just this boring yellow ball in the sky, there are many things to see on its surface, including **prominences**, **filaments**, flares and sunspots. These phenomena result from the strong magnetism within the Sun, which erupts to the surface.

Prominences are solar plasma ejections, some of which fall back to the Sun in the form of teardrops or loops; against the Sun's disk, prominences viewed from above appear as dark filaments. Sunspots are cooler regions of the Sun that appear dark on the face of the surrounding surface. If the highly magnetic lines of sunspots intertwine, they can snap and cause a solar flare — a bright, sudden eruption of energy that can last from a few minutes to several hours. It should be noted that to see prominences and some other solar features, you need very specialized filters.

If you do not have the proper equipment to observe the Sun, you can always visit the Solar and Heliospheric Observatory (SOHO) and the Solar Dynamics Observatory (SDO) websites (see the list of resources on page 128) to see daily images of the Sun.

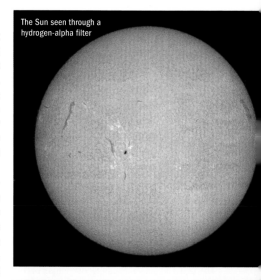

The Sun seen through a hydrogen-alpha filter

Sun with prominences, filaments, flares and sunspots

Eclipses

Eclipses — whether they're solar or lunar — are seemingly magical occurrences. In fact, that's exactly what many ancient civilizations believed. Several legends suggested a celestial being devoured the Sun during a total solar eclipse. For the Chinese, that being was either a dog or a dragon. The Vikings believed it was two wolves called Hati and Skoll. For the Vietnamese, it was a giant frog. Today, we understand that eclipses are awe-inspiring celestial events resulting from the movement and positions of the Earth, Moon and Sun. There can be as many as seven combined eclipses (i.e. solar and lunar) in a year, but no fewer than four.

A solar eclipse is truly a chance occurrence. The Sun's diameter is roughly 400 times that of the Moon, but the Sun is also about 400 times farther away from Earth than the Moon is from Earth. This means the Moon and the Sun appear roughly the same size in the sky, with only about 1/10th of a degree across of difference. In a **total solar eclipse**, the new Moon covers the entire face of the Sun, revealing the Sun's stunning corona and prominences. As totality begins and ends, sunlight peeking around the mountainous limb of the Moon creates fleeting effects such as the Diamond Ring and Baily's Beads. If the Moon were farther away or smaller, we wouldn't get this marvelous sight. In fact, the elliptical nature of both the Earth's orbit around the Sun and the Moon's orbit around the Earth means that the apparent diameters

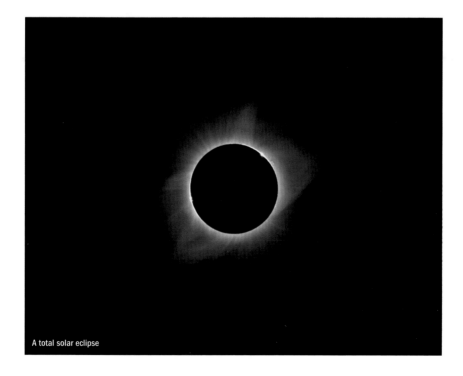

A total solar eclipse

of both objects vary significantly. Consequently, it is common for eclipse durations to vary by several minutes.

It should be noted that the totality only occurs in a very narrow path. Outside of this path, observers will only see a partial eclipse. That's why solar eclipse enthusiasts will travel great distances to take in the captivating sight of a totality. Remember, do not look at a solar eclipse unless you use special equipment recommended by The Royal Astronomical Society of Canada or the American Astronomical Society. The only time it is safe to look at a solar eclipse with the naked eye is when it is at its total phase, with the Sun completely covered by the Moon.

Because the Moon orbits at a 5-degree inclination to the ecliptic, we don't get a solar or lunar eclipse every month. Every year, there are two 36-day eclipse seasons during which either solar or lunar eclipses can occur, and the eclipse seasons move backward in time every year by 19 days. Also, because the Moon's orbit is elliptical, the angular size of the Moon varies along its path, and sometimes we have an **annular solar eclipse** (or "ring of fire"), when the Moon's disk does not completely cover the Sun's disk. There may be anywhere between zero to two total or annular solar eclipses in a given year. There may also be **partial solar eclipses**, when the Moon's central shadow entirely misses the Earth.

As for lunar eclipses, these occur when Earth is situated directly between the Sun and the full Moon. The Moon drifts through Earth's two shadows: the **penumbra**, which is the fainter, outer shadow, and the deeper shadow that is called the **umbra**. A **penumbral eclipse** is difficult to see with the naked eye as the brightness of the Moon doesn't appear to dim much. But a **partial** or **total lunar eclipse**, when the Moon passes through the umbra, is much more dramatic. During a total lunar eclipse, the Moon can turn an orange-red color, depending on Earth's atmosphere. There may be anywhere between zero to three total lunar eclipses in a given year. Typically, they occur about two weeks before or after a solar eclipse.

While not all eclipses will be visible from North America, you can watch online on sites like SLOOH or The Virtual Telescope Project. You can also see astronomer Fred Espenak's webpage eclipsewise.com for a comprehensive treatment of solar and lunar eclipses.

Eclipses in 2024

This year is a big eclipse year for North America with a total solar eclipse sweeping across the continent. There are sure to be viewing parties in Canada and the U.S., so you should keep an eye out for events near you. There's also a partial lunar eclipse that will be seen around the Atlantic Ocean region from Antarctica all the way to parts of the Arctic.

All the times listed are in UTC, or Coordinated Universal Time. See page 46 to learn how to calculate your local time from the time shown in UTC. Be sure to use a proper solar filter when

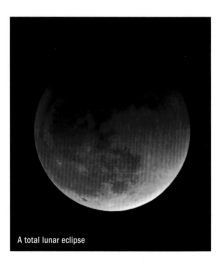

A total lunar eclipse

you are viewing a solar eclipse in person. Many of these events will also be live streamed online.

Lunar Eclipses

March 25, 2024: Penumbral Lunar Eclipse

Penumbral eclipses occur when the Moon passes through Earth's outer shadow, called the penumbra. Unfortunately, you likely won't notice much of a difference in the Moon's brightness as these eclipses are less dramatic than either total or partial lunar eclipses.

Eclipse times (UTC)	
Moon enters penumbra (P1)	04:53
Greatest eclipse	07:12
Moon exits penumbra (P4)	09:33

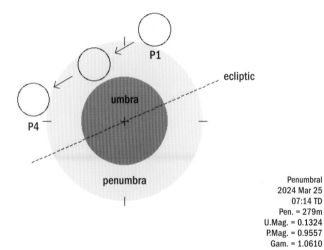

Penumbral
2024 Mar 25
07:14 TD
Pen. = 279m
U.Mag. = 0.1324
P.Mag. = 0.9557
Gam. = 1.0610

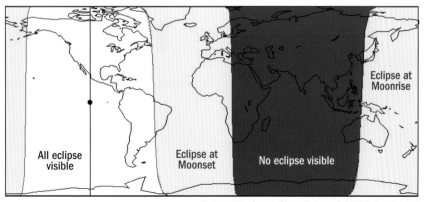

Thousand Year Canon of Lunar Eclipses © 2014 by Fred Espenak

September 18, 2024: Partial Lunar Eclipse

While this is considered a partial lunar eclipse, at most only 8.7 percent of the Moon's diameter will dip into the darkest part of Earth's shadow, the umbra. The rest of the event will be a penumbral lunar eclipse, when it will be difficult to notice much of a change in the color or brightness on the rest of the Moon. It might be a fun exercise to see how much you can notice, and binoculars may help.

Eclipse times (UTC)	
Partial eclipse begins (U1)	02:14
Greatest eclipse	02:42
Partial eclipse ends (U4)	03:17

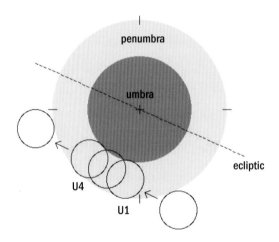

Partial
2024 Sept 18
02:45 TD
Par. = 63m
Gam. = -0.9792
U.Mag. = 0.0849
P.Mag. = 1.0373

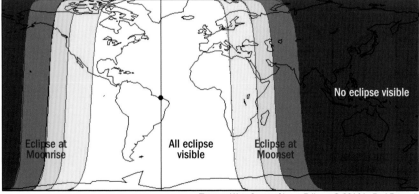

Thousand Year Canon of Lunar Eclipses © 2014 by Fred Espenak

Solar Eclipses

April 8, 2024: Total Solar Eclipse

Seven years after North America's last total solar eclipse, we will enjoy yet another. The 180-kilometer-wide (112-mile-wide) total eclipse track will begin in the Pacific Ocean, then cross Mexico, the United States, and eastern Canada and end in the North Atlantic Ocean 3.25 hours later. This eclipse is considered long, as totality will last almost four-and-a-half minutes. The partially eclipsed Sun will be visible across North and Central America, but not Alaska. See page 72 to learn more about this eclipse and totality viewing times across North America.

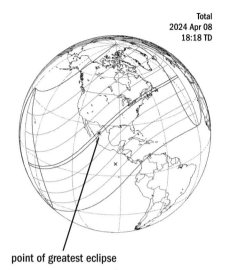

Total
2024 Apr 08
18:18 TD

point of greatest eclipse

Thousand Year Canon of Solar Eclipses
© 2014 by Fred Espenak

October 2, 2024: Annular Solar Eclipse

An annular solar eclipse occurs when the Moon orbits too far from Earth to cover the Sun entirely, leaving a ring of exposed Sun. This eclipse will mainly be visible over the southern Pacific Ocean and parts of Chile, Argentina, Paraguay, Uruguay and southern Brazil in the late afternoon.

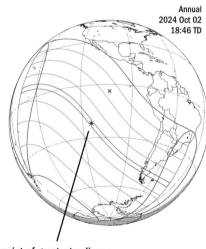

Annual
2024 Oct 02
18:46 TD

point of greatest eclipse

Thousand Year Canon of Solar Eclipses
© 2014 by Fred Espenak

The Northern Lights

The Sun appears to be a bright, unchanging orb in the sky. However, the Sun is anything but unchanging: As we know, it is a ball of constant activity, and that activity has a large influence here on Earth.

One such activity is a solar flare, which is often followed by a coronal mass ejection (CME) that sends particles speeding along the solar wind. Solar flares can reach Earth and disrupt radio transmissions. If Earth is in the path of a CME, the particles can travel down our magnetic field lines toward the poles; this creates the beautiful Northern and Southern Lights, or **aurora borealis** and **aurora australis**, respectively.

As mentioned earlier, the Sun goes through an average 11-year cycle of activity during which it experiences a solar minimum and a solar maximum. In the latter part of 2020, we began a new Solar Cycle, and as the cycle continues, we can expect to see more of the Northern Lights. It's worth noting that over the past few cycles the Sun has been less active than in the past.

Auroras come in different shapes and sizes. They can be steady, moving or rapidly pulsating. They also come in an array of colors, depending on how the particles interact with molecules at different altitudes. They are also hard to predict and can be difficult to see. Catching them is a special treat, even for experienced astronomers. Be aware that to the unaided eye the auroral colors are usually muted, as the light is not strong enough to stimulate our color vision. The bright colors seen in photographic reproductions of auroras show up when long exposures are used.

Green is the most common color and occurs when particles interact with oxygen molecules at an altitude of roughly 100 to 300 kilometers (62 to 186 miles). Between 300 to 400 kilometers (186 to 248 miles) the oxygen molecules produce a red color, and below 100 kilometers (62 miles) the particles interact with nitrogen — producing a pink color.

While the interaction of these solar particles can produce a beautiful light show, they can also be destructive.

One of the most powerful events from a CME was called the Carrington Event. In 1859, English astronomers Richard Carrington and Richard Hodgson were the first to ever witness a solar flare. But when the particles reached Earth, they disrupted telegraph systems in North America and Europe, in some reports even setting equipment on fire. The Northern Lights were visible as far south as Honolulu and the Southern Lights as far north as Santiago, Chile.

A massive CME, like the one that caused the Carrington Event, has a real chance of disrupting GPS and satellite communications and destroying electrical grids. As a result, space agencies have been launching more satellites and probes to the Sun to monitor space weather, and power companies have been developing contingency plans to ensure the next powerful solar eruption doesn't cause such damage.

The Northern Lights over Saskatchewan

The Planets

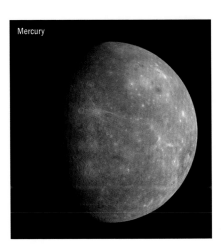

Mercury

As the planets (including Earth) revolve around the Sun, there are a number of notable events that take place, each of which has a specific name and meaning.

The inner planets Mercury and Venus never stray far from the Sun. When either planet is farthest from the Sun in the western evening sky, it is said to be at **greatest eastern elongation**; when either is farthest from the Sun in the eastern morning sky, it is said to be at **greatest western elongation**. These are the best opportunities to view the inner planets, though Venus can be seen at other times.

The term **conjunction** has a specific technical meaning — that is, when two objects have the same right ascension — but amateur astronomers also use the term casually when there is a close approach of two or more celestial objects.

Then there are **oppositions**: when the outer planets (Mars, Jupiter, Saturn, Uranus, Neptune) are opposite the Sun in the sky in right ascension. Around opposition dates, the outer planets appear closer, brighter and larger in diameter — great for telescopic viewing. The Sky Month-by-Month section (pages 46–121) lists eastern and western elongations, conjunctions and oppositions throughout 2024.

Mercury is a place of extremes. As the smallest of our Solar System's eight planets at one-third the size of Earth, Mercury is the closest planet to the Sun, sitting an average distance of 58 million kilometers (36 million miles) away. The planet, however, is only the second-hottest one in our Solar System. Mercury also orbits the Sun faster than any other planet and has the longest solar day, lasting 176 Earth days. Mercury has no moon.

The planet's surface warms to 427 degrees Celsius (801 degrees Fahrenheit) at the peak of its "noonday" heat; on the far side, its temperature drops down to a chilly –163 degrees Celsius (–261 degrees Fahrenheit). Mercury's coldest locations are deep in the shadows of the craters, where the Mercury Surface, Space Environment, Geochemistry and Ranging (MESSENGER) spacecraft mission to the small planet discovered ice. Temperatures in crater shadows are at –183 degrees Celsius (–297 degrees Fahrenheit).

Due to the planet's proximity to the Sun, Mercury is a challenging target to observe. When it is visible in Earth's sky, it appears in a narrow window of time in the dawn twilight just before sunrise or in the evening twilight just after sunset, depending on whether it is west or east of the Sun in the sky. To observe Mercury, one needs a low horizon unobstructed by trees or buildings, as it is typically only 15 to 25 degrees away from the Sun when visible.

Venus, the brightest planet in the sky, is a spectacular sight to see. Known as both the "morning star" and the "evening star," Venus is similar in size to Earth. But it's definitely not a place you'd like to visit. The planet is covered in dense clouds and, as a result, suffers from a runaway greenhouse effect. The planet's surface has temperatures of 460 degrees Celsius (860 degrees Fahrenheit) almost constantly.

A view of the Martian landscape from NASA's Perseverance rover in March 2021

Venus spins in the opposite direction than the other planets, but very slowly. A day on Venus is longer than its year: one day takes 243 Earth days, while one year is 225 Earth days. Its orbit is, on average, 108 million kilometers (67 million miles) from the Sun. Being shrouded in white clouds, there is nothing much to see on Venus, but binoculars and small telescopes reveal that Venus passes through phases, similar to the Moon. (Mercury also shows phases, but they are more difficult to see, as the planet's disk appears so small.)

In 2024, Venus can be found in the east just before sunrise in late March and early April. In late March it will be joined by Saturn.

There is no other planet that fascinates humans as much as **Mars**. In 1894, American astronomer Percival Lowell thought he could see deep canals carved through its surface that he believed were evidence of intelligent life. Later observations with spacecraft saw no evidence of these canals.

Mars is the most explored planet in our Solar System, and some even like to joke that it's the only planet with an entirely robotic "civilization" of Earth-made machines.

The orbiters, landers and rovers dispatched to the Red Planet have taught us a lot about its ancient past. Though rocky, dusty and seemingly devoid of any life, some evidence suggests Mars once had an ocean of water covering most of its northern hemisphere. Scientists sometimes cite the existence of water to suggest the planet was once habitable, though we have yet to find proof that life actually ever thrived there.

Mars is about half the size of Earth, with a day very close to Earth's 24-hour day. It lies roughly 228 million kilometers (142 million miles) away from the Sun. Small telescopes (with a 70 mm diameter and below) will show the red disk of Mars; a larger, good-quality telescope at high magnification will show surface features such as ice caps, dark basins and light deserts.

In 2024, Mars and Saturn can be found in the eastern sky just before sunrise beginning in March. Mars will remain there until August, when it will be joined by Jupiter. There will be an early morning conjunction of the pair on August 14, where they will appear to almost touch.

Jupiter is the king of our planetary system, at 11 times wider than Earth. This gas giant boasts the most spectacular storm formation in our Solar System, called the Great Red Spot. Jupiter orbits 772 million kilometers (479 million miles) from the Sun and a single day is about 10 Earth hours long. It orbits the Sun in about 12 Earth years.

Jupiter is the second-brightest planet in our night sky and is also one of the most enjoyable planets to observe. With a modest telescope, you can easily see the cloud bands in its atmosphere. But you can even enjoy the planet through a pair of binoculars; if you watch the planet every night, four of its 92 confirmed moons — Ganymede, Io, Europa and Callisto — can be seen changing positions night after night.

The Great Red Spot (GRS) is a massive, swirling egg-shaped storm that has been shrinking for reasons that are unclear to astronomers. In the 1800s, the GRS was estimated at 41,000 kilometers (25,476 miles) along its long axis. NASA's Juno spacecraft measured the GRS at 16,350 kilometers (10,159 miles) in width on April 3, 2017. Several astronomy apps predict the best time on any night for viewing the GRS through a telescope.

Jupiter

With its magnificent rings, **Saturn** is often considered the jewel of our Solar System. With a pair of binoculars, Saturn's pancake shape is evident. Look through a telescope, even a modest one, and its fine rings are clearly visible.

Saturn is nine times wider than Earth and 1.4 billion kilometers (886 million miles) away from the Sun. This gaseous planet has the second-shortest day in the Solar System, lasting roughly only 10.7 hours, while its orbit takes 29.4 Earth years. The planet also has more than 80 moons, though its largest, Titan, is the only one easily viewed through a modest telescope.

Saturn is the second-largest planet in the Solar System. It's not the only planet to have rings (Jupiter, Uranus and Neptune all have ring systems), but it has by far the most spectacular ring system we can see. Saturn's rings are made up of billions of pieces of ice that range from tiny dust-sized grains to chunks as big as a house. The ring system is roughly 282,000 kilometers (175,000 miles) across but only about 9 meters (30 feet) tall. There are several rings, mostly close to each other, but the main rings are called A, B, and C. The gap between the outer ring (A) and the first inner ring (B) is called the Cassini Division, following the discovery by Italian astronomer Giovanni Cassini.

Uranus is the faintest of the planets visible to the naked eye (in dark-sky conditions), and the third-largest planet overall. This giant ice planet has 13 faint rings and 27 small moons. But most interestingly, Uranus rotates on its side, at nearly a 90-degree angle.

In 1781, Uranus was the first planet to be discovered with the aid of a telescope by astronomer William Herschel. However, Herschel first believed it was either a star or a comet. It was confirmed as a planet two years later by Johann Elert Bode.

The planet is four times the size of Earth. A day on Uranus takes roughly 17 Earth hours, with one orbit taking 84 Earth years. Uranus

Saturn

orbits on average 2.9 billion kilometers (1.8 billion miles) from the Sun.

This year, Uranus can be found in Taurus. It will reach opposition on November 16, where it will be at a magnitude of +5.6.

At an average distance of 4.5 billion kilometers (2.8 billion miles) from the Sun, **Neptune** is the farthest planet in our Solar System, taking about 165 Earth years to complete one orbit. Like Uranus, Neptune is roughly four times larger than Earth, and it, too, has a faint ring system.

Neptune also holds the distinction as being the windiest planet in our Solar System, with clouds of frozen methane being whipped across the planet at 2,000 kilometers (1,200 miles) per hour. A day on Neptune is about 16 Earth hours long.

Neptune will reach opposition on September 20, where it will reach a magnitude of +7.8.

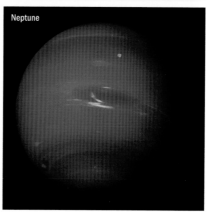

Neptune

Deep-Sky Objects

The Universe has many amazing sights to behold, including stunning clusters of stars, nebulae and swirling galaxies.

There are two main types of star clusters: **open** and **globular**. Globular clusters are old star systems at the edge of spiral galaxies that can contain anywhere from thousands to millions of stars, packed in a close, roughly spherical form and held together by gravity.

Two beautiful globular clusters you can see from the Northern Hemisphere include Messier 13, found in the constellation Hercules (a hero from Greek mythology), and Messier 3 in the constellation Canes Venatici (the Hunting Dogs).

Open clusters are found on the **galactic plane**, the plane on which most of a galaxy's mass lies. They contain anything from a dozen to hundreds of stars, but the stars are more spread out than globular clusters. Perhaps the most famous open cluster, the Pleiades (Messier 45) can be spotted with the naked eye and through light pollution.

As well, there are **nebulae** — clouds of dust and gas. These are considered "stellar nurseries," as eventually the gas and dust will coalesce into new stars and potential stellar systems with planets, moons and, possibly, life. One of the most famous and easily visible is the Orion Nebula (Messier 42), found in the winter constellation Orion.

Some nebulae are leftover dust and gas from a **supernova**, an exploding star. There are also **emission nebulae**, which are clouds of interstellar gas excited by nearby stars that emit their own light at optical wavelengths. **Planetary nebulae** are cloudy remnants left over from stars that shed their gas and dust late in their lives. And finally, there are **dark nebulae**, interstellar clouds that are so dense they obscure the light of the objects behind them.

Messier 13, a globular cluster, seen here with Mars (at far left)

Messier Catalog

The Messier Catalog is a register of 110 objects in the night sky, including open and globular clusters, nebulae, galaxies and one double star. The catalog was started by Charles Messier in the 18th century. Messier was chiefly interested in finding comets, and the catalog is his list of non-comet objects that he observed. You can find a list of all the objects in the Messier Catalog on pages 122–125.

Pleiades (Messier 45)

Orion Nebula (Messier 42)

Galaxies

Galaxies come in many different shapes and sizes. There are four main types, however: elliptical, spiral, barred spiral and irregular. The diagram below shows Edwin Hubble's scheme for classifying galaxies, colloquially known as the "tuning fork" because of its shape.

Elliptical galaxies seem somewhat disorganized and look roughly egg-shaped. Shapes range from almost circular (E0) to very elliptical (E7).

Spiral galaxies have both a large central bulge and a thin disk of stars. Much like elliptical galaxies, these also range from tight spirals (Sa) to more diffuse (Sd).

Barred spiral galaxies are similar to spirals, but they have visible "arms" or bars near the center, and they range from tight (SBa) to diffuse (SBd).

There are also disk galaxies that don't have spiral arms. These are called **lenticular (lens-shaped) galaxies**. They are classified as S0.

The most common galaxy is the spiral, accounting for more than 75 percent of galaxies in the visible Universe. Our galaxy, the Milky Way, is believed to be a barred spiral. Our closest neighboring spiral galaxy is the Andromeda Galaxy (Messier 31), which is easily visible through binoculars or with the naked eye in dark skies.

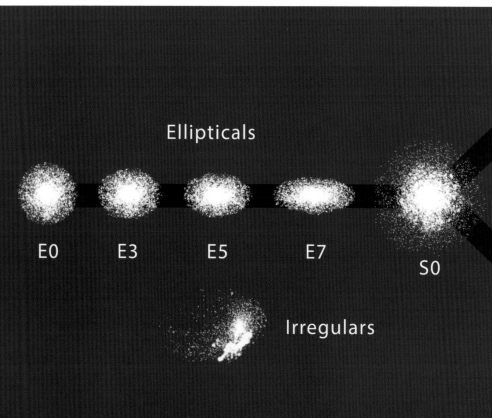

Ellipticals

E0   E3   E5   E7   S0

Irregulars

Andromeda Galaxy (Messier 31)

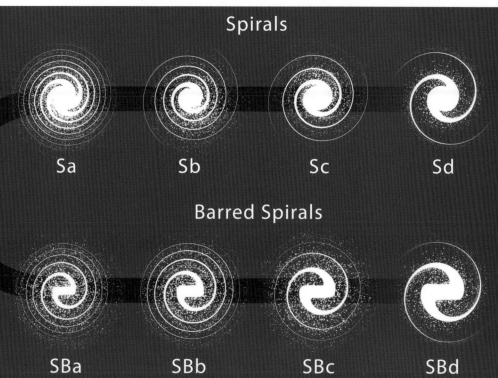

Spirals

Sa   Sb   Sc   Sd

Barred Spirals

SBa   SBb   SBc   SBd

The Sky Month-by-Month

Introduction

The following pages are your guide to the sky for every month in 2024. Each month features a calendar of events, information about Moon phases, sky charts facing south and north and descriptions of interesting objects to target.

Monthly Events

This section summarizes the events of each month, including Moon phases, conjunctions between the Moon and planets, conjunctions between planets and other planets, eastern and western elongations, meteor shower peaks and much more.

All times are listed in UTC, or Coordinated Universal Time, which is the standard time used by astronomers throughout the year. The map shown below will guide you on how to calculate your local time from the time shown in UTC. It's important to note that most of North America will observe Daylight Saving Time (DST) between March 10, 2024, and November 3, 2024. Between these two dates, you will need to add an additional hour to your local time.

You'll notice some of the times listed fall during daylight, when observation is likely impossible. However, you can use this date

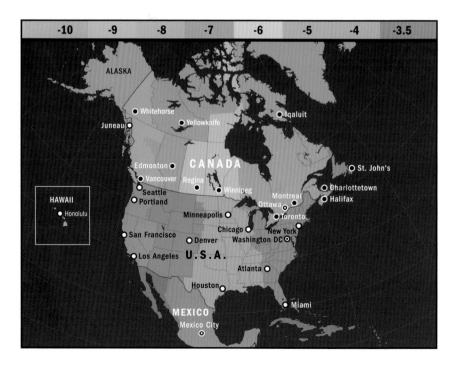

and time as a guide for observing an event on either the preceding night or the following night. Depending on your location, objects might also be below the horizon at the stated time, so again, use this calendar as a guide for which night these events can be observed. Similarly, the coordinates given as the distance between celestial objects may not reflect exactly what you will see in the night sky. This is particularly true for close approaches that take place during the day, local time. For example, the calendar of events might say Mercury is 1.5 degrees south of Jupiter at 21:00 UTC, but by the time Mercury and Jupiter become visible to an observer in Eastern Time living in Toronto, Canada, that separation might have increased to 2.4 degrees.

This section also features a calendar layout with the Moon phases for each day shown, as well as a description of interesting Moon events for the month. You might notice some discrepancies between the coordinates given in these descriptions and those listed in the table of events. Because the Moon moves in right ascension so quickly, conjunctions between the Moon and planets as listed in the table may appear different for North American observers, and the separation between the objects may be more than suggested. The coordinates in the descriptions have been altered to more closely reflect the view from North America, but of course there may be more variation based on when the objects are visible to you in the night sky.

The Moon phases shown in the calendar are based on the day they occur in UTC time. For some observers in North America, that might mean the ideal time to watch, for example, the full Moon rising would be the night before.

Sky Charts and Descriptions

Each month features two sky charts, one facing south and the other facing north. These charts highlight select constellations, stars, clusters, nebulae, planets and galaxies visible in the night sky. While these sky charts are an accurate representation of the sky, it should be noted that factors like light pollution, smoke, cloud cover and so on can affect how much you see, and some of the fainter stars shown on the charts may not be visible.

The charts are drawn at a latitude of 45 degrees north. Observers north or south of this latitude will see slightly more of the northern or southern sky, respectively. The charts show the sky at 10:00 p.m. local standard time on the 15th of each month. You will also have the same view of the sky at 11:00 p.m. local standard time at the beginning of each month and at 9:00 p.m. local standard time at the end of each month. We note these times on the sky charts and also include Daylight Saving Time in parentheses between March and November. The planets shown on the sky charts will move slightly relative to the stars over the course of the month.

The preceding month's charts can be used for sky viewing two hours earlier, and the following month's charts can be used for sky viewing two hours later. So, for example, if you wanted to view the sky at 8:00 p.m. local time in February, you would refer to January's sky chart. If you're referring to a previous or later month's sky chart, note that the positions of the planets will not be correct.

January

January's Events

The new year has been rung in, and it's time for another year of great astronomical objects to enjoy.

The month starts off with the Quadrantids, which peaks in the early morning of January 4. This meteor shower has the potential to be one of the best of the year, but with the peak lasting only a few hours, it can be tricky to view. Though the Quadrantids are capable of producing 100-plus meteors an hour under ideal conditions, most observers will likely see only a handful. But the good news is that the few you do see are likely to be impressive fireballs.

Jupiter will be the planet dominating the January night sky, lying in Aries.

Calendar of Events

Day	Time (UTC)	Event
01	15:28	Moon at apogee: 404,900 km (251,593 mi.)
02	23:59	Earth at perihelion: 0.9833 astronomical units (au)
04	3:30	Last quarter of the Moon
04		Quadrantid meteor shower peak
08	20:12	Venus 5.9°N of Moon
11	11:57	New Moon
12	13:59	Mercury 23.5°W of Sun (greatest western elongation)
13	10:35	Moon at perigee: 362,300 km (225,123 mi.)
14	9:31	Saturn 2.1°N of Moon
18	3:53	First quarter of the Moon
18	20:40	Jupiter 2.9°S of Moon
20	13:25	Pleiades 0.9°N of Moon
24	19:00	Pollux 1.9°N of Moon
25	17:54	Full Moon
27	15:48	Mars 0.2°N of Mercury
29	8:14	Moon at apogee: 405,800 km (252,152 mi.)

The Moon This Month

SUN	MON	TUES	WED	THURS	FRI	SAT
	1	2	3	4 Last Quarter	5	6
7	8	9	10	11 New Moon	12	13
14	15	16	17	18 1st Quarter	19	20
21	22	23	24	25 Full Moon	26	27
28	29	30	31			

The Quadrantids peak on the night of January 3 to 4, which is good news, as the Moon will only be roughly five days old, allowing observers the chance to potentially catch some of the fainter meteors.

In the morning sky throughout January, observers are treated to the beautiful Venus shining brightly in the southeast before sunrise. Alongside the "morning star," is Mercury, which will reach its highest point in the sky on January 7.

On January 8, in the eastern morning sky, a waning crescent Moon will join the pair of planets, sitting just southwest of Venus, close to the bright star Antares.

On January 14, just after sunset, the Moon and Saturn will lie roughly 5 degrees apart in the southwest. And on January 18, the Moon and Jupiter have their close encounter in the night sky, lying only 3 degrees apart.

On January 20, the Moon swings by the beautiful jewel box of the Pleiades (Messier 45), which will make an excellent binocular target.

The new Moon falls on January 11 and the full Moon on January 25.

The Moon reaches apogee (the farthest from Earth in its monthly orbit) both on January 1 and 29 and it obtains perigee (the closest to Earth in its monthly orbit) on January 13.

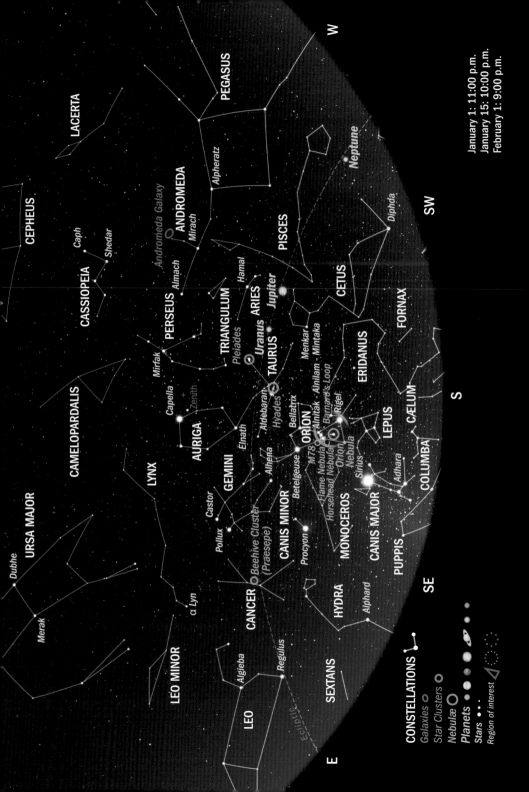

Highlights in the Southern Sky

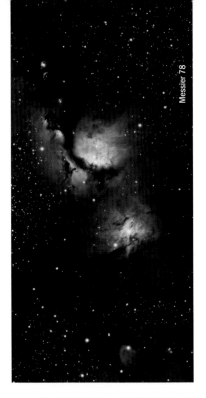

Messier 78

The winter months are the favorites for astronomers since the nights are longer — even if they have to contend with clouds and cold temperatures. But one of the great treats on a clear winter's night is the mighty **Orion (the Hunter)**. The constellation is probably the most recognizable in the sky, marked by its three belt stars, from left to right, **Alnitak, Alnilam** and **Mintaka**.

What makes this constellation truly remarkable is that it lies in a region of our galaxy loaded with dust and nebulae. There is truly a wealth of objects to see either through binoculars or even small telescopes. The most prominent, of course, is the **Orion Nebula (Messier 42)**, which lies just below the belt stars and is meant to represent Orion's sword. This magnificent star-forming region is even visible to the unaided eye as a fuzzy "star."

There's also **Barnard's Loop (Sh 2-276)**, a half-moon–shaped area of faint red nebulosity that is most spectacular in photographs.

Other amazing Orion targets are the **Horsehead Nebula (Barnard 33)** and the **Flame Nebula (NGC 2024)**. And let's not forget **Messier 78.** Though it's not as well-known as some of its other Orion cousins, this reflection nebula (one that is illuminated by a nearby star) can be seen in large binoculars as a fuzzy patch.

Another great thing about Orion is its stars. One of the most well-known is **Betelgeuse**, a red supergiant that someday (perhaps hundreds of thousands of years from now, or sooner) will explode into a brilliant supernova.

South of Betelgeuse and the belt stars is the brilliant blue supergiant **Rigel**, the brightest star in Orion. Even though it looks like one star, it is a system of several.

If you've looked toward Orion, you'll have undoubtedly noticed a particularly brilliant star to the east of the belt stars. That is **Sirius**, or the "Dog Star," the brightest star in our sky and part of the constellation **Canis Major (the Great Dog)**.

The winter sky also holds two of the best open star clusters: the **Hyades (Melotte 25)** and the **Pleiades (Messier 45)**. You can find the Hyades by a nearby prominent star, **Aldebaran** in **Taurus (the Bull)** — the Hyades help form the bull's head and horns. The Pleiades will sit roughly 14 degrees from Aldebaran.

Another beautiful star cluster, the **Beehive Cluster (Messier 44)**, is located 50 degrees to the left of Orion. Also known as **Praesepe (the Manger)**, it is one of the closest clusters to Earth, lying just 600 light-years away. The cluster contains about 1,000 stars that are a mere 600 million years old. (Remember: Our own Sun in 4.5 billion years old.)

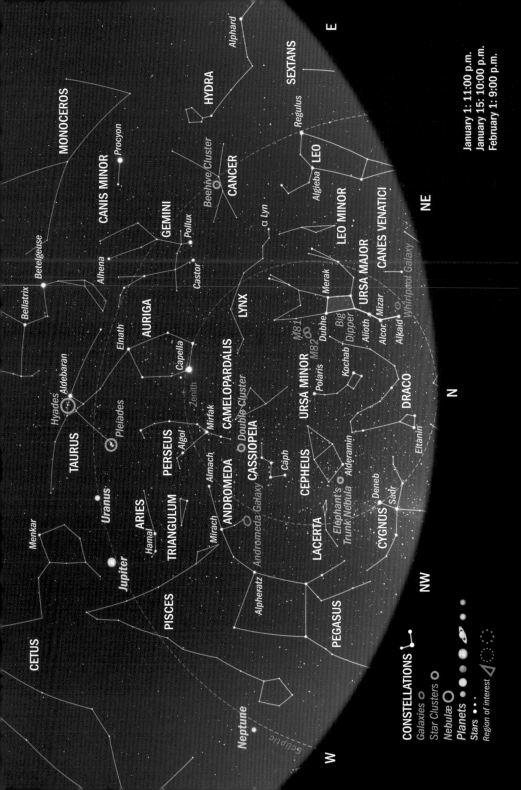

E

MONOCEROS

HYDRA

Alphard

SEXTANS

CANIS MINOR

Procyon

GEMINI

Beehive Cluster

CANCER

Regulus

LEO

Betelgeuse

Pollux

Alhena

α Lyn

Algieba

LEO MINOR

NE

Bellatrix

Castor

AURIGA

LYNX

URSA MAJOR

CANES VENATICI

Elnath

Capella

CAMELOPARDALIS

Merak

Dubhe

Big
Dipper

Whirlpool Galaxy

Aldebaran

Zenith

Double Cluster

M81

Alioth

Hyades

Mirfak

M82

Alcor

Mizar

Alkaid

Plejades

Algol

PERSEUS

CASSIOPEIA

URSA MINOR

Kochab

TAURUS

Almach

Caph

Polaris

DRACO

ANDROMEDA

Eltanin

ARIES

Uranus

Mirach

CEPHEUS

Hamal

TRIANGULUM

Andromeda Galaxy

Alderamin

Jupiter

Elephant's
Trunk Nebula

N

PISCES

LACERTA

Deneb

Alpheratz

CYGNUS

Sadr

Menkar

PEGASUS

NW

CETUS

Neptune

Ecliptic

W

January 1: 11:00 p.m.
January 15: 10:00 p.m.
February 1: 9:00 p.m.

CONSTELLATIONS

Galaxies ⊙

Star Clusters ⊙

Nebulæ ⊙

Planets ● ● ● ●

Stars ● ● · ·

Region of interest

Highlights in the Northern Sky

Messier 81.

Turning your eyes to the northeast you can see what may be the most recognizable asterism of the Northern Hemisphere, the **Big Dipper**, which also resembles a giant spoon. If you have keen eyes and look carefully at the second star of the spoon's handle, you may notice that it's actually made up of two stars, **Alcor** and **Mizar** (Mizar being the brighter of the pair). Binoculars will show their separation well. Some stories say ancient civilizations, including the Romans and the Arabs, would use the pair as a visual acuity test. In January, you can find the Big Dipper standing upright on its handle, with the bottom of the bowl pointing to the left. The asterism belongs to **Ursa Major (the Great Bear)**, which can be used to find the celestial north pole marked by **Polaris**, also known as the **North Star**. Polaris belongs to **Ursa Minor (the Little Bear)**, which is also often referred to as the **Little Dipper**.

To find Polaris using the Big Dipper as a guide, find the two stars at the top of the bowl, **Merak** to the right and **Dubhe** to the left. Trace a line from Merak to Dubhe. If you keep going along the same line, it will lead you to Polaris. The star appears almost stationary in the sky night after night, month after month, in the Northern Hemisphere. If you were to set up a camera pointing directly at Polaris, you'd notice the constellations circling around it as the night progresses.

If it's galaxies you love, you're in luck this month. One of the most photographed galaxies in the region is the **Whirlpool Galaxy (Messier 51a)**, which can be found near **Alkaid**, the star at the end of the Big Dipper's handle. This galaxy can be seen through a modest telescope.

Two wonderful galaxies also lie within Ursa Major: **Messier 81 (Bode's Galaxy)** and **Messier 82 (the Cigar Galaxy)**. Messier 81 can be seen at dark-sky sites through binoculars or a small telescope, and Messier 82

is in the same field of view. The two are often photographed together as they are extremely close in the night sky.

February

February's Events

The very bright Venus will spend February sinking lower and lower in the southeastern morning sky before sunrise. Mercury is now invisible as it is too near the Sun. On February 21, Venus will pass close to Mars. If you're hoping to catch the pair, you'll need a clear view of the eastern horizon. By the end of the month, the morning sky washes out Venus altogether. In the evening, Jupiter is still well placed for viewing in Aries.

Orion still dominates the sky in the evening. It is well-positioned for viewing in the south. Taurus sits to the right of Orion, with the bright red star Aldebaran shining brightly at the lower left corner of the Hyades open cluster. (That star is not actually considered to be part of the cluster.) Nearby, another beautiful open star cluster, the Pleiades, adorns the sky. Binoculars will reveal myriad stars in both.

Orion

Calendar of Events

Day	Time (UTC)	Event
02	23:18	Last quarter of the Moon
07	18:52	Venus 5.4°N of Moon
08	6:30	Mars 4.2°N of Moon
09	22:59	New Moon
10	18:49	Moon at perigee: 358,100 km (222,513 mi.)
15	8:15	Jupiter 3.2°S of Moon
16	15:01	First quarter of the Moon
16	19:13	Pleiades 0.6°N of Moon
22	9:01	Mars 0.6°N of Venus
24	12:30	Full Moon
25	15:00	Moon at apogee: 406,300 km (252,463 mi.)

The Moon This Month

SUN	MON	TUES	WED	THURS	FRI	SAT
				1	2 Last Quarter	3
4	5	6	7	8	9 New Moon	10
11	12	13	14	15	16 1st Quarter	17
18	19	20	21	22	23	24 Full Moon
25	26	27	28	29		

The Moon with Jupiter overhead

On February 2, the Moon will be in its last quarter, with a new Moon on February 9. The first quarter of the Moon occurs on February 16, with the full Moon dominating the sky on February 24.

On February 14, you can find a lovely waxing crescent Moon less than 5 degrees west of the Solar System's largest planet, Jupiter.

The Moon will be at perigee on February 10 and at apogee on February 25.

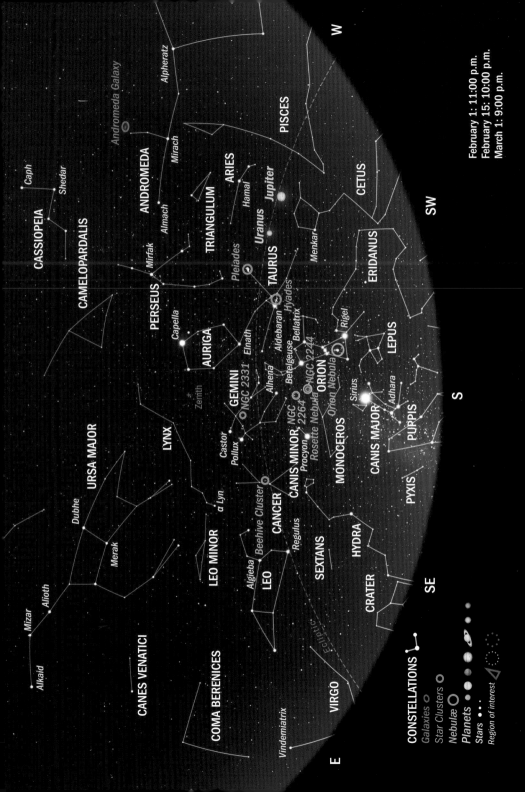

Cone Nebula

Highlights in the Southern Sky

Orion (the Hunter) continues to be the most remarkable constellation in the southern sky, but there are other fantastic treats to enjoy.

One such constellation is **Gemini (the Twins)**. The constellation lies just to the left of Orion's sword and is marked by its two brightest stars **Pollux** and **Castor**, which represent the heads of the mythological Greek twins. Castor is just 51 light-years distant and is part of a multiple-star system that includes three binary star pairs, meaning there is a total of six stars in all. Pollux is an orange giant that lies a bit closer to Earth at 34 light-years away. The open cluster **NGC 2331** in central Gemini is visible with the aid of binoculars.

Just south of Gemini and to the east of Orion is **Monoceros (the Unicorn)**, a somewhat faint constellation. It is home to the beautiful **Rosette Nebula**, a star-forming region that lies 5,200 light-years from Earth. At its heart is a rich cluster of stars called **NGC 2244.**

Another treat in Monoceros is the **Cone Nebula**, which also includes a beautiful star-forming region called the **Christmas Tree Cluster** (the pair are known collectively as **NGC 2264**). While the stars are fairly easy to spot with binoculars, a telescope is needed to see the Cone Nebula.

Jupiter can be found easily as it is the brightest object in the south-western sky, lying in **Aries (the Ram)**.

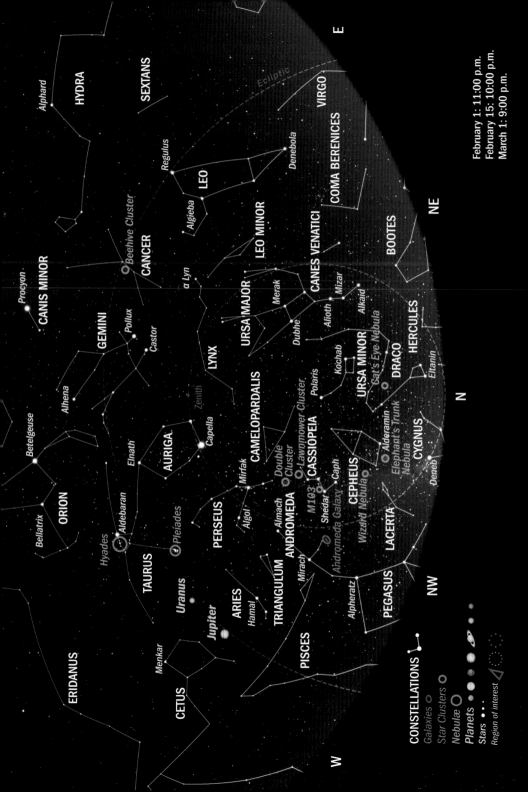

Pacman Nebula

Highlights in the Northern Sky

Cassiopeia (the Queen), which looks like a sideways W, is well placed in the northwest at the beginning of February but begins to sink lower throughout the month.

The constellation — named after the boastful and vain queen of Greek legend — is a treasure trove of objects, including the **Lawnmower Cluster (NGC 663)** and the nearby cluster **Messier 103**. If you use a pair of binoculars to scan the region, you won't be disappointed with the number of stars you'll see.

The Milky Way that descends through Cassiopeia down to the horizon is a region full of clusters and some beautiful nebulae, such as the **Pacman Nebula (NGC 281)**, the **Wizard Nebula (NGC 7830)** and the **Elephant's Trunk Nebula (IC 1396)** located in **Cepheus (the King)** — all are favorite targets for astrophotographers.

Near Cassiopeia is the constellation **Perseus (the Hero)**, home to two big open clusters known as the **Double Cluster (NGC 869 and NGC 884)**. They are easily visible even from moderately dark locations and make fantastic binocular targets.

In the north lies **Draco (the Dragon)** winding northwest and then northeast before its tail stretches north, past **Ursa Minor (the Little Bear)**. Draco is home to a planetary nebula called the **Cat's Eye Nebula (NGC 6543)**, which is believed to be expanding at a rate of 10.2 miles (16.4 kilometers) per second. Planetary nebulae are the leftovers from dying stars that shed their gas and dust near the end of their lives. Once our own Sun consumes all its hydrogen in a few billion years, it may create one of these beautiful stellar remnants.

March

March's Events

By now, you may have noticed the later sunsets making the nights shorter. As well, the more familiar constellations like Orion are beginning to sink farther and farther west as the night progresses. If you're an early riser, turn to the southeast before sunrise and you may notice summer constellations like Sagittarius and Scorpius starting to rise.

March also marks an important month for astronomers. This month, amateur astronomers like to complete the Messier Marathon. If you're interested, it's probably good to know that no athletic abilities are required, just keen eyes.

Astronomers try to "chase" the 110 Messier objects (from Charles Messier's Catalog) using binoculars or telescopes. The goal? To view all of them on a single night. If you're interested, a pair of binoculars with an aperture of 70 millimeters or greater would do the job. You can find a list of the Messier objects on pages 122-125.

When it comes to planets, sadly, there aren't many in the sky this month. Jupiter remains visible but can be found low in the west.

The vernal equinox occurs on March 20, marking the beginning of spring.

Calendar of Events

Day	Time (UTC)	Event
03	15:24	Last quarter of the Moon
08	4:59	Mars 3.5°N of Moon
08	17:01	Venus 3.2°N of Moon
10	7:06	Moon at perigee: 356,900 km (221,767 mi.)
10	9:00	New Moon
14	1:01	Jupiter 3.6°S of Moon
17	4:11	First quarter of the Moon
20	3:07	Vernal (spring) equinox
21	22:06	Saturn 0.3°S of Venus
23	15:44	Moon at apogee: 406,300 km (252,463 mi.)
24	21:59	Mercury 18.7°E of Sun (greatest elongation east)
25	7:00	Full Moon
25	7:13	Penumbral Lunar Eclipse

The Moon This Month

SUN	MON	TUES	WED	THURS	FRI	SAT
					1	2
3 Last Quarter	4	5	6	7	8	9
10 New Moon	11	12	13	14	15	16
17 1st Quarter	18	19	20	21	22	23
24	25 Full Moon	26	27	28	29	30
31						

On March 3, the Moon will be at last quarter, with the new Moon occurring just seven days later, on March 10, the same day it is at perigee.

On March 14, the waxing crescent Moon will skirt the edge of the Pleiades.

The first quarter is on March 17 and the full Moon falls on March 25. Also on March 25 is a penumbral lunar eclipse. Unfortunately, you likely won't notice much of a difference in the Moon's brightness as penumbral eclipses are less dramatic than either total or partial lunar eclipses.

Pleiades

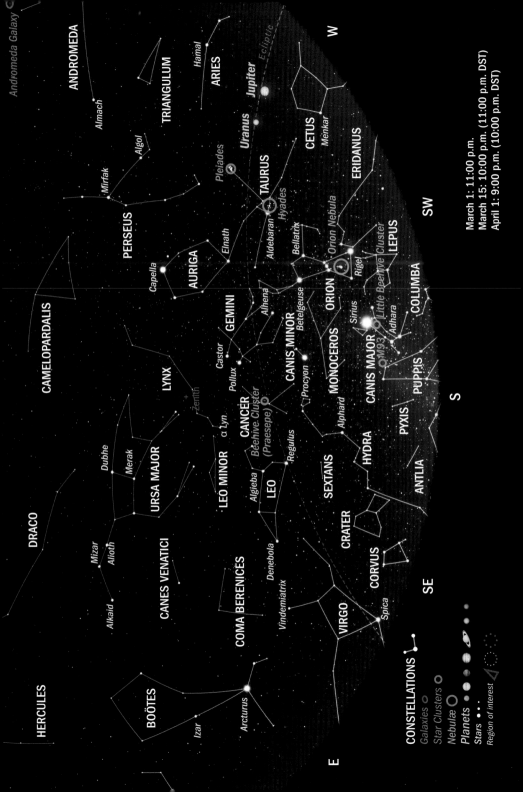

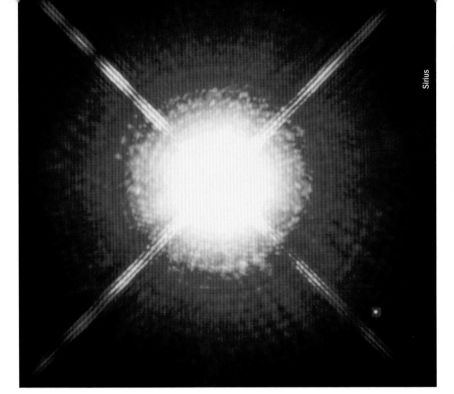

Highlights in the Southern Sky

You may now notice an incredibly bright star in the south. That would be **Sirius**, the most prominent star in **Canis Major (the Great Dog)**, just south of **Monoceros (the Unicorn)**. While Canis Major never climbs too high in the sky, you can find beautiful binocular and telescopic targets near it, including the **Little Beehive Cluster (Messier 41)**.

Nearby is the open cluster **Messier 93**, found in the southerly constellation **Puppis (the Poop Deck)**. The cluster lies just 3,600 light-years from Earth and, while it can be seen with the unaided eye under dark skies, it makes a fantastic binocular target.

To the southeast you can find **Hydra (the Water Snake)**, the longest of the 88 constellations. Just above it is **Cancer (the Crab)**, where binoculars will reveal a stunning open cluster — home to roughly 1,000 stars — the **Beehive Cluster (Messier 44)**, which is also known as **Praesepe** or the **Manger**. It is just 600 light-years away, making it one of the closest open clusters to Earth.

To the southwest, in **Taurus (the Bull)**, you'll find the beautiful open cluster the **Hyades (Melotte 25)** along with the **Pleiades (Messier 45)**.

By the end of the month, **Orion (the Hunter)** clearly dominates the southwestern sky, and **Leo (the Lion)** sits high in the south.

CRATER

VIRGO

Vindemiatrix

Arcturus

Izar

CORONA BOREALIS

COMA BERENICES

Coma Cluster

CANES VENATICI

BOÖTES

M13

HERCULES

Denebola

LEO

Regulus

Algieba

SEXTANS

LEO MINOR

Alkaid

Mizar

Alioth

Merak

Dubhe

URSA MAJOR

DRACO

Eltanin

CANCER

Beehive Cluster

α *Lyn*

Kochab

Polaris

URSA MINOR

Alderamin

μ *Cephei*
(Garnet Star)

HYDRA

LYNX

CAMELOPARDALIS

CASSIOPEIA

CEPHEUS

Caph

Shedar

LACERTA

Procyon

CANIS MINOR

Pollux

Castor

GEMINI

Alhena

AURIGA

Capella

Elnath

Mirfak

PERSEUS

Algol

TRIANGULUM

Almach

Andromeda Galaxy

ANDROMEDA

Triangulum Galaxy

MONOCEROS

Betelgeuse

Bellatrix

ORION

Orion Nebula

Rigel

LEPUS

ERIDANUS

Aldebaran

Hyades

TAURUS

Pleiades

Uranus

Jupiter

ARIES

Hamal

Mirach

PISCES

CETUS

Zenith

Ecliptic

E

NE

N

NW

W

March 1: 11:00 p.m.
March 15: 10:00 p.m. (11:00 p.m. DST)
April 1: 9:00 p.m. (10:00 p.m. DST)

CONSTELLATIONS
Galaxies ○
Star Clusters ○
Nebulæ ○
Planets ●
Stars • • •
Region of interest △

Highlights in the Northern Sky

One of the most prominent stars in the Northern Hemisphere is **Arcturus** in the constellation **Boötes (the Herdsman)**, which is now beginning to rise in the northeast. Arcturus — the brightest in the constellation — is an interesting star, as it is a red giant that's roughly 25 times larger than our Sun, and it is relatively close to Earth at a distance of 37 light-years. The constellation itself looks somewhat like a kite.

To the north is **Cepheus (the King)**, resembling a little house. Cepheus is another great binocular target and is also quite special for its "jewel." At the base of the house-shaped constellation is **Mu (μ) Cephei**, or the "Garnet Star," given its name because of its deep reddish color. The star is variable, meaning that it fluctuates in brightness.

Those with a telescope may want to turn it toward **Canes Venatici (the Hunting Dogs)**, **Coma Berenices (Berenice's Hair)** and **Virgo (the Maiden)**. Within this region lies thousands of galaxies belonging to the **Coma Cluster**, which lies roughly 330 million light years away and spans about 25 million light-years. This, in turn, is believed to be part of an even bigger supercluster of galaxies that spans a mind-blowing 200 million light-years.

Be sure to keep an eye out for the tiny constellation **Triangulum (the Triangle)**. It is home to the **Triangulum Galaxy (Messier 33)**, the third-largest member of the Local Group of galaxies that includes the **Andromeda Galaxy (Messier 31)** and our own **Milky Way**. The Triangulum Galaxy lies just 3 million light-years away from Earth. It's a tough target to see with the unaided eye, but at a dark-sky location it can be seen through a pair of binoculars. It is visible in telescopes.

Triangulum Galaxy

March 2024

April

April's Events

This April is a very special month for North Americans as viewers will be treated to a total solar eclipse on April 8, the first one since 2017. The eclipse path will cross from northern Mexico, across the central U.S. and into eastern Canada (see pages 72–73 for more details).

The winter constellations are moving out and the spring constellations are riding high. If you're an early riser, you'll be able to catch some of the summer constellations beginning to make their appearance. On the mornings of both April 10 and 11, early risers will also be treated to a beautiful conjunction of Mars and Saturn in the eastern sky, where they will be less than half a degree apart. The pair will slowly separate over the following days.

There's also the Lyrid meteor shower to look forward to this month. Though not a strong shower (usually it has a ZHR of 18), it is known for producing fireballs. The peak falls on the night of April 21 to 22, but a nearly full Moon will somewhat spoil the show.

Calendar of Events

Day	Time (UTC)	Event
02	3:15	Last quarter of the Moon
06	3:51	Mars 2°N of Moon
06	9:20	Saturn 1.2°N of Moon
07	16:39	Venus 0.4°S of Moon
07	17:53	Moon at perigee: 358,800 km (222,948 mi.)
08	18:18	Total solar eclipse
08	18:21	New Moon
10	18:46	Saturn 0.4°N of Mars
10	21:08	Jupiter 4°S of Moon
15	13:47	Pollux 1.6°N of Moon
15	19:13	First quarter of the Moon
20	2:09	Moon at apogee: 405,600 km (252,028 mi.)
22		Lyrid meteor shower peak
23	23:49	Full Moon

The Moon This Month

SUN	MON	TUES	WED	THURS	FRI	SAT
	1	2 Last Quarter	3	4	5	6
7	8 New Moon	9	10	11	12	13
14	15 1st Quarter	16	17	18	19	20
21	22	23 Full Moon	24	25	26	27
28	29	30				

On April 6, a thin crescent Moon will lie just below Saturn with Mars nearby.

On the night of April 23 to 24, there is a conjunction with an almost-full Moon and the star Spica, found in Virgo.

The last quarter falls on April 2 and the new Moon on April 8 (the reason why we get a total solar eclipse, with the Moon in front of the Sun). The first quarter is on April 15, with the full Moon occurring on April 23.

The Moon is at perigee on April 7 and at apogee on April 20.

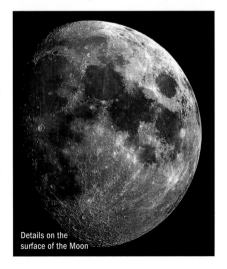

Details on the surface of the Moon

Markarian's Chain

Highlights in the Southern Sky

Virgo (the Maiden) can now be seen in the southeast. This second-largest constellation is home to one of the most intriguing galaxies in our sky, the **Sombrero Galaxy (Messier 104)**. Also contained in the region is the **Virgo Cluster** of roughly 2,000 galaxies, and it is the closest cluster to the Milky Way. One of the most notable grouping of galaxies is **Markarian's Chain**, a favorite object for many amateur astronomers and astrophotographers.

Just above is **Coma Berenices (Berenice's Hair)**. While not overly prominent, this constellation is also home to another collection of 1,000 galaxies, called the **Coma Cluster**.

Boötes (the Herdsman) is now much higher in the sky with its shining star **Arcturus** an easy-to-spot target.

Looking to the southwest, you can find **Leo (the Lion)** and its two brightest stars, **Regulus** and **Denebola**. Just south of Leo are three galaxies grouped together called the **Leo Triplet**, which contains **Messier 65, Messier 66** and **NGC 3628** (referred to as the **Hamburger Galaxy**).

To the west is **Cancer (the Crab)**, home to the **Beehive Cluster (Messier 44)**. And to its west is **Gemini (the Twins)**, which is now beginning to descend toward the western horizon.

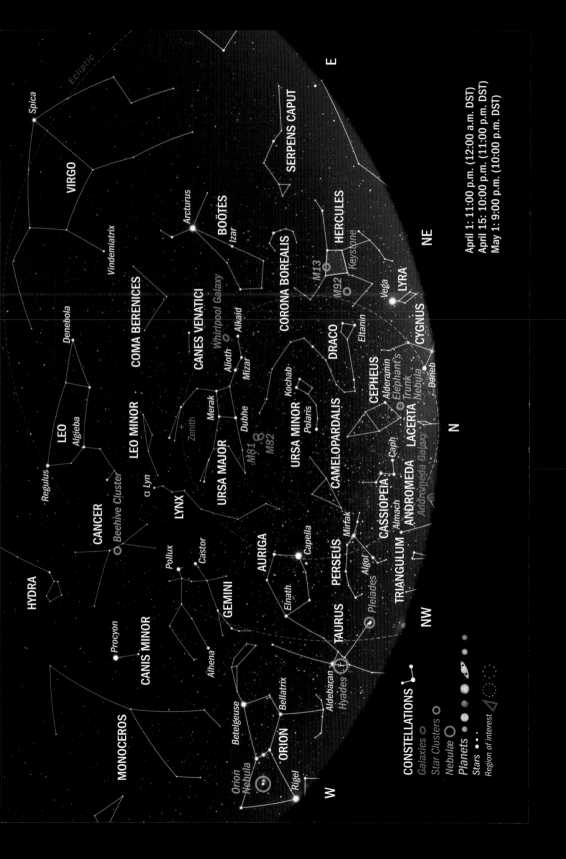

E

April 1: 11:00 p.m. (12:00 a.m. DST)
April 15: 10:00 p.m. (11:00 p.m. DST)
May 1: 9:00 p.m. (10:00 p.m. DST)

NE

N

Ecliptic

Spica

VIRGO

Arcturus

BOÖTES

Vindemiatrix

Izar

SERPENS CAPUT

CORONA BOREALIS

Denebola

COMA BERENICES

HERCULES

M13
M92
Keystone

Whirlpool Galaxy

CANES VENATICI

Alkaid
Alioth
Mizar

Vega

LYRA

Regulus

LEO

Algieba

LEO MINOR

Beehive Cluster

Merak

Dubhe

Kochab

DRACO

Eltanin

CYGNUS

CANCER

α Lyn

Zenith

URSA MAJOR

M81
M82

URSA MINOR

Polaris

CAMELOPARDALIS

CEPHEUS

Alderamin

Elephant's
Trunk
Nebula

Deneb

HYDRA

Procyon

LYNX

CASSIOPEIA

Caph

LACERTA

ANDROMEDA

Andromeda Galaxy

CANIS MINOR

Pollux

Castor

GEMINI

AURIGA

Capella

PERSEUS

Mirfak

Algol

TRIANGULUM

Almach

Alhena

Elnath

Pleiades

NW

Betelgeuse

MONOCEROS

Bellatrix

ORION

Aldebaran
Hyades

TAURUS

Orion
Nebula

Rigel

W

CONSTELLATIONS

Galaxies

Star Clusters

Nebulæ

Planets

Stars

Region of interest

Highlights in the Northern Sky

Hercules is now rising in the northeast, marked by its familiar "keystone," which is home to one of the most beautiful globular clusters in the Northern Hemisphere, **Messier 13**. It contains roughly 100,000 stars. Under dark-sky conditions — and with keen eyes — this cluster can be seen without the aid of binoculars. The cluster was discovered by Sir Edmund Halley, who is best known for the comet he discovered, Halley's Comet. Nearby you can also find **Messier 92**, another remarkable cluster that is best viewed through binoculars or a small telescope.

If it's galaxies you like, this is a great time to turn your eyes high overhead to **Ursa Major (the Great Bear)**, where you can spot **Messier 81 (Bode's Galaxy)** and **Messier 82 (the Cigar Galaxy)**. Messier 81 can be seen under dark skies without the aid of binoculars, while a telescope will reveal the pair in the same field of view.

Meanwhile, in **Canes Venatici (the Hunting Dogs)**, you can find the **Whirlpool Galaxy (Messier 51)**, another favorite of amateur astronomers and astrophotographers. The best way to find it is by locating the star **Alkaid** in Ursa Major. This is a remarkable spiral galaxy located roughly 23 million light-years away.

Whirlpool Galaxy.

The Great North American Eclipse

On April 8, 2024, sky watchers in North America will be treated to a total solar eclipse — the first in seven years. The total eclipse begins in the Pacific Ocean, then it crosses Mexico, the United States, Canada and finally the North Atlantic Ocean. This eclipse is considered long, as totality will last almost four-and-a-half minutes. The partially eclipsed Sun will be visible across North and Central America.

When it comes to viewing solar eclipses, remember that safety comes first. Do not look at a solar eclipse unless you use special equipment recommended by The Royal Astronomical Society of Canada or the American Astronomical Society. The only time it is safe to look at a solar eclipse with the naked eye is when it is at its total phase, the brief period of time when the Sun is completely covered by the Moon.

This handy table shows you when totality begins and ends in a city in each U.S. state and Canadian province that intersects with the path of totality. The times are listed in local time.

Location	Partial Begins	Totality Begins	Maximum	Totality Ends	Partial Ends
Dallas, Texas	12:23 CDT	13:40 CDT	13:42 CDT	13:44 CDT	15:02 CDT
Idabel, Oklahoma	12:28 CDT	13:45 CDT	13:47 CDT	13:49 CDT	15:06 CDT
Little Rock, Arkansas	12:33 CDT	13:51 CDT	13:52 CDT	13:54 CDT	15:11 CDT
Poplar Bluff, Missouri	12:39 CDT	13:56 CDT	13:56 CDT	14:00 CDT	15:15 CDT
Paducah, Kentucky	12:42 CDT	14:00 CDT	14:01 CDT	14:02 CDT	15:18 CDT
Evansville, Indiana	12:45 CDT	14:02 CDT	14:04. CDT	14:05 CDT	15:20 CDT
Cleveland, Ohio	13:59 EDT	15:13 EDT	15:15 EDT	15:17 EDT	16:29 EDT
Erie, Pennsylvania	14:02 EDT	15:16 EDT	15:18 EDT	15:20 EDT	16:30 EDT
Buffalo, New York	14:04 EDT	15:18 EDT	15:20 EDT	15:22 EDT	16:32 EDT
Burlington, Vermont	14:14 EDT	15:26 EDT	15:27 EDT	15:29 EDT	16:37 EDT
Lancaster, New Hampshire	14:16 EDT	15:27 EDT	15:29 EDT	15:30 EDT	16:38 EDT
Caribou, Maine	14:22 EDT	15:32 EDT	15:33 EDT	15:34 EDT	16:40 EDT
Toronto, Ontario	14:04 EDT	N/A	15:19 EDT	N/A	16:31 EDT
Ottawa, Ontario	14:11 EDT	N/A	15:25 EDT	N/A	16:35 EDT
Montreal, Quebec	14:14 EDT	15:26 EDT	15:27 EDT	15:28 EDT	16:36 EDT
Fredericton, New Brunswick	15:33 ADT	16:33 ADT	16:34 ADT	16:36 ADT	17:42 ADT

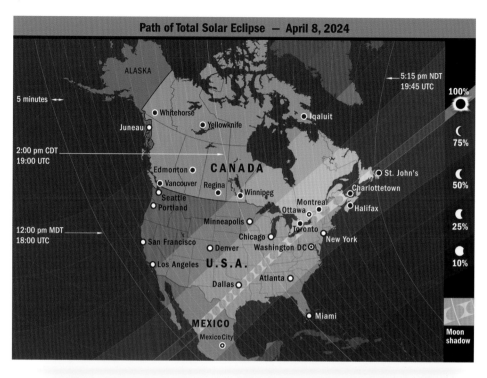

Path of Total Solar Eclipse — April 8, 2024

ALASKA

5 minutes

Whitehorse

Juneau

Yellowknife

Iqaluit

5:15 pm NDT
19:45 UTC

100%

75%

50%

25%

10%

2:00 pm CDT
19:00 UTC

Edmonton

CANADA

Vancouver   Regina

Winnipeg

St. John's

Charlottetown

Seattle

Portland

Montreal

Ottawa

Halifax

12:00 pm MDT
18:00 UTC

Minneapolis

San Francisco

Denver

Chicago

Toronto

New York

Washington DC

Los Angeles   U.S.A.

Atlanta

Dallas

Moon
shadow

MEXICO

Miami

Mexico City

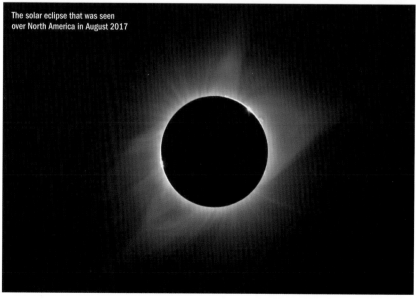

The solar eclipse that was seen
over North America in August 2017

May

May's Events

Daylight hours are now longer, which is bad news for stargazers. Still, it's a good time to begin planning out your summer night-sky viewing for those vacation days you may be able to spend away from the glare of city lights. If you're unable to get to a cottage, just try to find a dark-sky location — a beach or a park — where you're able to enjoy the night sky as the warmer weather begins to descend upon us.

Mars and Saturn are now beginning to rise in the morning sky in the east. So, if you're an early bird, keep an eye out because by the end of the month they will be lost in the glare of the rising Sun.

As well, May gives us the annual Eta Aquarid meteor shower, which actually began in mid-April. The shower is best viewed from the Southern Hemisphere, but you may be able to catch a few meteors on the moonless peak night of May 4 to 5. The ZHR is roughly 10 to 30 meteors an hour.

Calendar of Events

Day	Time (UTC)	Event
01	11:27	Last quarter of the Moon
03	22:26	Saturn 0.8°N of Moon
04		Eta Aquarid meteor shower
05	2:26	Mars 0.2°N of Moon
05	22:11	Moon at perigee: 363,200 km (225,682 mi.)
06	8:25	Mercury 3.8°S of Moon
08	3:22	New Moon
09	20:59	Mercury 26.4°W of Sun (greatest elongation west)
15	11:48	First quarter of the Moon
17	19:00	Moon at apogee: 404,600 km (251,407 mi.)
23	13:53	Full Moon
30	17:13	Last quarter of the Moon
31	8:01	Saturn 0.4°N of Moon

The Moon This Month

SUN	MON	TUES	WED	THURS	FRI	SAT
			1 Last Quarter	2	3	4
5	6	7	8 New Moon	9	10	11
12	13	14	15 1st Quarter	16	17	18
19	20	21	22	23 Full Moon	24	25
26	27	28	29	30 Last Quarter	31	

On May 1, we get the last quarter of the Moon, with the new Moon falling on May 8. The full Moon occurs on May 23 and the last quarter on May 30.

On May 4, a waning crescent Moon will lie between Mars and Saturn in the early hours before sunrise, very low on the eastern horizon.

On May 31, grab a pair of binoculars, as there will be a conjunction of the waning crescent Moon and Saturn.

The Moon is at perigee on May 5 and at apogee on May 17.

The dark areas of the Moon can be faintly illuminated by earthshine, when the Sun's light reflects off the Earth's surface

Messier 5

Highlights in the Southern Sky

As **Virgo (the Maiden)** continues to dominate the southern sky, you can find **Libra (the Scales)** beginning to rise in the southeastern sky. It is one of the zodiac constellations and the only one that represents an object rather than an animal or person.

Near Libra is **Serpens Caput (the Head of the Serpent)**, where you can find **Messier 5**, also known as the **Rose Cluster**, a beautiful globular cluster that lies 24,500 light-years from Earth and is best seen in binoculars and telescopes.

Hydra (the Water Snake) is now low on the southern horizon, slinking its way along to the west. The largest constellation in the night sky, it is home to a few Messier objects, including the globular cluster **Messier 68**, which can be seen through large binoculars but is best seen in a telescope, even a small one. Also within Hydra is **Messier 48**, an open cluster that can be spotted without the aid of binoculars under dark-sky conditions but is best seen through binoculars and small telescopes.

North of Hydra are the small constellations **Corvus (the Crow)** and **Crater (the Cup)**. While neither hold any particularly bright stars, Corvus is home to a pair of interacting galaxies called the **Antennae Galaxies (NGC 4038 and 4039)**. It's believed that roughly four billion years from now, our Milky Way and the **Andromeda Galaxy (Messier 31)** will similarly collide.

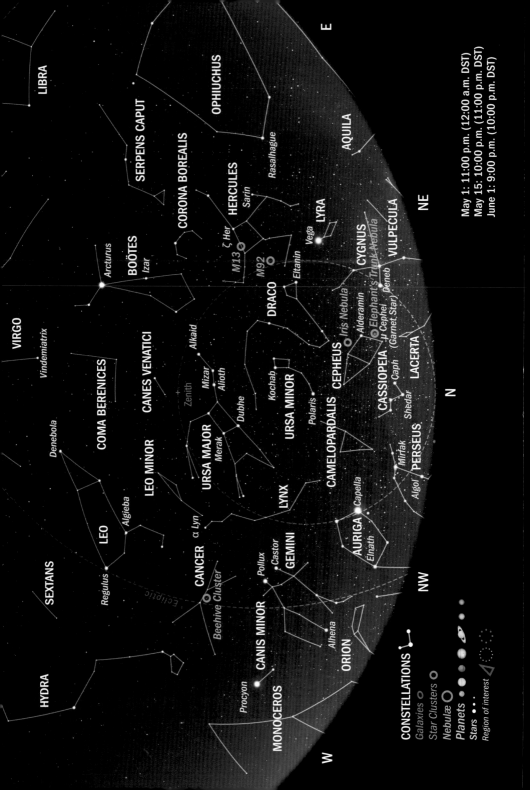

Elephant's Trunk Nebula

Highlights in the Northern Sky

Cepheus (the King) is now beginning to rise higher in the north. Some particularly notable nebulae can be found in the constellation, including a favorite of many astrophotographers, the **Iris Nebula (NGC 7023)**, with its bright blue-purple flower-like core surrounded by intense gas and dust.

Another great photo subject is the **Elephant's Trunk Nebula**. This is actually part of a much bigger star-forming region of gas and dust called **IC 1396.**

But one of the main highlights in Cepheus is the "Garnet Star," also known as **Mu (μ) Cephei**. This bright red supergiant may be one of the largest naked-eye stars, with a radius of roughly four billion kilometers (2.49 billion miles). If it sat at the center of our Solar System, it would extend beyond the orbit of Saturn.

Cassiopeia (the Queen) is beginning to sink lower in the north, while **Cygnus (the Swan)** starts to rise in the northeast. **Lyra (the Lyre)** is also rising, with its star **Vega**, the fifth-brightest star in the sky, easily visible.

The mighty **Hercules** is now high in the northeast, where you can find two beautiful globular clusters, **Messier 13** and **Messier 92**.

June

June's Events

The rich summer constellations, such as Scorpius, Sagittarius and Ophiuchus, are now beginning to rise higher in the sky. These constellations are home to some of the most beautiful objects, including numerous clusters and nebulae.

As well, Cygnus, Lyra and Aquila are now above the horizon in the east, each of their brightest stars making up a well-known asterism called the Summer Triangle. This region — especially around Cygnus — is home to many bright clusters and other targets that are a favorite of many amateur astronomers.

At the beginning of the month, Mars can be found very low in the east just before sunrise.

The summer solstice falls on June 20.

The Milky Way with the Summer Triangle asterism

Calendar of Events

Day	Time (UTC)	Event
02	7:23	Moon at perigee: 368,100 km (228,727 mi.)
02	23:37	Mars 2.4°S of Moon
06	12:38	New Moon
14	5:18	First quarter of the Moon
14	13:36	Moon at apogee: 404,100 km (251,096 mi.)
20	20:51	Summer solstice
22	1:08	Full Moon
27	11:45	Moon at perigee: 369,300 km (229,472 mi.)
27	14:52	Saturn 0.1°S of Moon
28	21:53	Last quarter of the Moon

The Moon This Month

SUN	MON	TUES	WED	THURS	FRI	SAT
						1
2	3	4	5	6 New Moon	7	8
9	10	11	12	13	14 1st Quarter	15
16	17	18	19	20	21	22 Full Moon
23	24	25	26	27	28 Last Quarter	29
30						

When it comes to lunar and planetary conjunctions, on June 2, a waning crescent Moon will be roughly 5 degrees from a dim Mars on the eastern horizon. On June 27, the Moon and Saturn will be separated by about 3 degrees before sunrise.

The new Moon falls on June 6, with the first quarter occurring on June 14. The full Moon is on June 22 and the last quarter is on June 28.

The Moon is at perigee twice this month, on June 2 and 27, and is at apogee on June 14.

No matter the time of year, the Moon always makes a wonderful target in the night sky

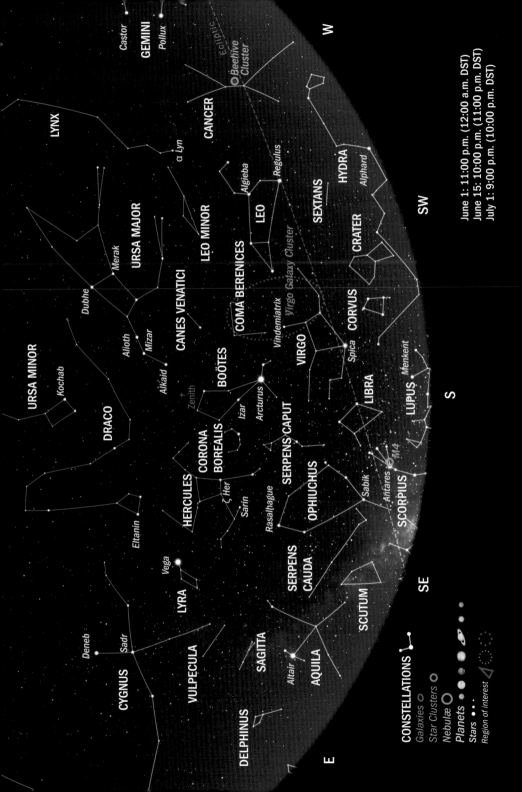

The summer Milky Way with Antares and Messier 4

Highlights in the Southern Sky

Ophiuchus (the Serpent Bearer) is now higher in the southern sky, separated by **Serpens Caput (the Head of the Serpent)** and **Serpens Cauda (the Tail of the Serpent)**. **Serpens (the Serpent)** is a unique constellation in that it is separated in two parts by the much larger Ophiuchus.

You may have noticed a bright red star in the south, under Ophiuchus, which would be **Antares (the Rival of Mars)**. Antares, part of the constellation **Scorpius (the Scorpion)**, is a massive red supergiant like Orion's Betelgeuse and is several hundred times larger than the diameter of our Sun – and roughly 10,000 times more luminous. Lying some 600 light-years away, Antares is nearing the end of its life. As a red supergiant star, it will eventually die in a spectacular explosion called a supernova. Antares also has a smaller, dimmer companion called **Antares B**, though it is hard to see next to its much bigger and brighter partner.

Very near Antares is **Messier 4**, a bright globular cluster that lies roughly 7,200 light-years from Earth, making it one of the closest clusters. **Virgo (the Maiden)** is still visible but is beginning to descend into the western horizon. **Coma Berenices (Berenice's Hair)** is also well placed. Both these constellations are home to thousands of galaxies in what are known as the **Virgo Cluster** and the **Coma Cluster**.

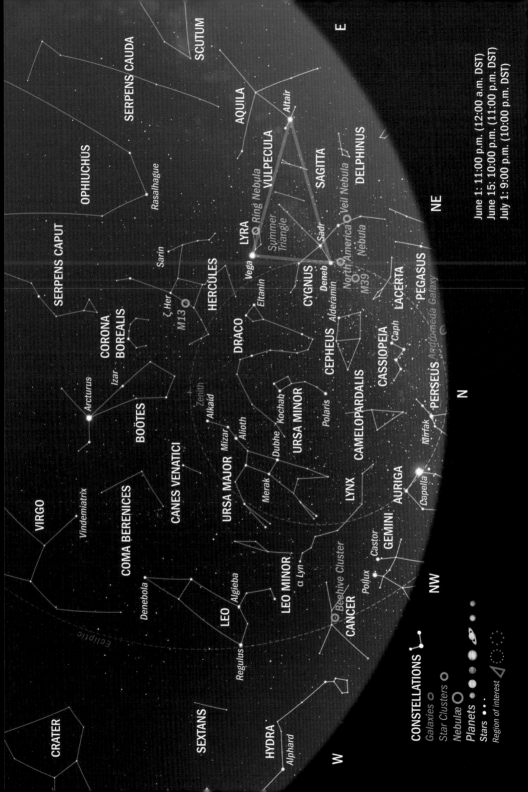

Highlights in the Northern Sky

Ursa Major (the Great Bear), with its well-known asterism the **Big Dipper**, is now very high in the sky. **Draco (the Dragon)** winds around nearby **Ursa Minor (the Little Bear)**.

Cygnus (the Swan) is flying higher in the northeast. Its brightest star is **Deneb**, which represents the tail of the swan. The star lies roughly 1,500 light-years away and is about 200 times more massive than our Sun. It is one of the most distant stars that we can see with the naked eye. Nearby is **Lyra (the Lyre)**, with its own bright star **Vega**. And finally, there is **Aquila (the Eagle)** with its bright star **Altair**. All three bright stars make up the popular asterism called the **Summer Triangle**.

Cygnus is a beautiful binocular target, rich with stars. There are several open clusters, including **Messier 39**, and it is also home to several well-known nebulae, such as the **North America Nebula (NGC 7000)** and the **Veil Nebula**, created by a supernova explosion. The region is large and is often captured by astrophotographers. But because it is so large, it is broken up into two regions: the **Eastern Veil Nebula (NGC 6992)** and the **Western Veil Nebula (NGC 6960)**, often referred to as the **Witch's Broom**.

A beautiful sight in Lyra is the **Ring Nebula (Messier 57)**, which looks just as its name suggests. It is roughly 2,000 light-years away and, though not visible with the unaided eye, its ring formation is clearly visible through small telescopes.

June is also a great month to try to observe **noctilucent clouds**, clouds that form high in the atmosphere and occur in the Northern Hemisphere beginning around the middle of May. While these clouds aren't entirely understood, it's believed that they are ice coalescing on the dust of meteoroids high up in our atmosphere. Decades ago, these iridescent clouds used to only be seen in polar regions, but more recently, they have been spotted farther south — even as far as Los Angeles, California.

Noctilucent clouds

July

July's Events

While the nights may be short now, they will start to get longer this month, by a minute or two per day. Though the hot summer nights may lack in length, they are the favorites of many stargazers. The Milky Way stretching magnificently overhead — from the densest areas lying in Sagittarius and Scorpius, to Scutum, Aquila and Cygnus. The best thing is that, for the most part, you only need your eyes or a pair of binoculars to enjoy the many targets available.

The next two months are a great time to look for star parties in your area. They are usually hosted by astronomical groups and offer a weekend to gaze up at the stars in dark-sky locations away from city lights. If you don't have a telescope, most of the amateur astronomers who attend are happy to let you peer through theirs as well as guide you to the best targets. There's truly a wealth of knowledge to obtain from others in your sky-watching community.

Jupiter is now visible in the eastern morning sky just before sunrise. Joining the mighty planet are Mars and Saturn higher in the south.

A summer star party

Calendar of Events

Day	Time (UTC)	Event
01	18:27	Mars 4.2°S of Moon
05	5:59	Earth at aphelion: 1.0167 au
05	22:57	New Moon
07	18:33	Mercury 3.5°S of Moon
12	8:12	Moon at apogee: 404,400 km (251,283 mi.)
13	22:49	First quarter of the Moon
21	10:17	Full Moon
22	06:59	Mercury 26.9°E of Sun (greatest eastern elongation)
24	5:43	Moon at perigee: 364,900 km (226,738 mi.)
24	20:38	Saturn 0.4°S of Moon
28	2:52	Last quarter of the Moon

The Moon This Month

SUN	MON	TUES	WED	THURS	FRI	SAT
	1	2	3	4	5 New Moon	6
7	8	9	10	11	12	13 1st Quarter
14	15	16	17	18	19	20
21 Full Moon	22	23	24	25	26	27
28 Last Quarter	29	30	31			

On July 1, Mars and the waning crescent Moon will be separated by roughly 4 degrees. On July 24 and 25, a gibbous Moon will swing by Saturn.

The new Moon falls on July 5 and the first quarter on July 13. The full Moon occurs on July 21, with the last quarter falling on July 28.

The Moon is at apogee on July 12 and at perigee on July 24.

The Moon with Mars close by

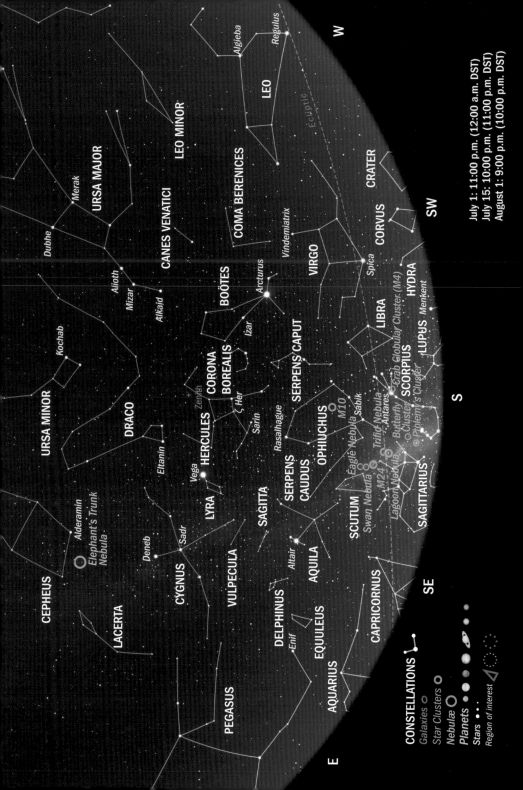

Trifid Nebula

Highlights in the Southern Sky

The southern sky provides a wealth of targets to see this month.

Sagittarius (the Archer) is relatively easy to spot in the south; just look for its teapot shape in the sky. One of the constellation's most notable features — which can be seen with the unaided eye, particularly at dark-sky locations — is the **Sagittarius Star Cloud (Messier 24)**. Found just north of the teapot's lid, this cloud is home to millions of stars, and is roughly 600 light-years wide.

A short hop above Messier 24 is the **Swan Nebula (Messier 17)**, which is also known as the **Omega Nebula,** and is clearly visible in binoculars and small telescopes. This star-forming nebula lies roughly 5,000 light-years from Earth and is part of a larger cloud.

Just north of the Swan Nebula is the **Eagle Nebula (Messier 16)**. This nebula became famous from an image taken by the Hubble Space Telescope in 1995 that NASA dubbed "The Pillars of Creation" due to its dense clouds of dust and gas, which are forming new stars.

Another great target is the **Lagoon Nebula (Messier 8)**, a bright star-forming region near the teapot's lid. Its cluster of stars is easily visible through binoculars. Close by it is the **Trifid Nebula (Messier 20)**, given its name due to its three-lobed appearance. There are several objects in the nebula, including a dark nebula and an open cluster.

In the next-door constellation **Scorpius (the Scorpion)** lie several open clusters, including **Ptolemy's Cluster (Messier 7)**. This can be found near the stinger of Scorpius and is fairly bright. Binoculars reveal dozens of beautiful stars. There is also the **Butterfly Cluster (Messier 6)**, which sits right near the center of our galaxy.

You can also find **Messier 10** in **Ophiuchus (the Serpent Bearer),** another beautiful globular cluster that also lies near the galactic center.

E

AQUARIUS

Enif

AQUILA

Altair

SERPENS CAUDA

OPHIUCHUS

Rasalhague

SERPENS CAPUT

EQUUELEUS

DELPHINUS

M15

PEGASUS

NE

VULPECULA

SAGITTA

Veil Nebula

Elephant's Trunk Nebula

NGC 7243

LACERTA

Andromeda Galaxy

LYRA

Deneb

Sadr

CYGNUS

CEPHEUS

CASSIOPEIA

ANDROMEDA

Vega

Alderamin

Caph

Shedar

Mirfak

Eltanin

CORONA BOREALIS

HERCULES

Sarin

ζ Her

M13

Zenith

DRACO

URSA MINOR

Kochab

Polaris

PERSEUS

Capella

N

Izar

Whirlpool
Galaxy Alkaid

Mizar

Alioth

Dubhe

CAMELOPARDALIS

BOÖTES

Arcturus

Merak

LYNX

Vindemiatrix

CANES VENATICI

URSA MAJOR

NW

COMA BERENICES

LEO MINOR

Algieba

α Lyn

VIRGO

Denebola

LEO

Regulus

Ecliptic

W

Spica

CONSTELLATIONS
Galaxies ○
Star Clusters ○
Nebulæ ○
Planets ● ● ● ● ●
Stars ● ● ·
Region of interest △

July 1: 11:00 p.m. (12:00 a.m. DST)
July 15: 10:00 p.m. (11:00 p.m. DST)
August 1: 9:00 p.m. (10:00 p.m. DST)

Pegasus Cluster

Highlights in the Northern Sky

Cassiopeia (the Queen) is now rising higher in the northeast. Binoculars will reveal several open clusters, as the region lies within a particularly busy region of the Milky Way.

Andromeda (the Princess) along with **Pegasus (the Flying Horse)** are beginning to appear in the northeast. Pegasus, marked by its obvious square shape, is where you can find the **Pegasus Cluster (Messier 15)**, a globular cluster very near the small, faint constellation **Equuleus (the Foal)**.

Delphinus (the Dolphin), a tiny kite-like constellation, lies just above Equuleus. Another small constellation called **Lacerta (the Lizard)** can be found between Pegasus and **Cepheus (the King)**, which is in the northern sky to the right of Polaris. Lacerta is also home to **NGC 7243**, a nice, though somewhat faint, open cluster visible through binoculars.

To the northwest, **Ursa Major (the Great Bear)** is beginning to sink lower. This may be a good time to check out the **Whirlpool Galaxy (Messier 51)**, which lies in **Canes Venatici (the Hunting Dogs)** but is located near the star **Alkaid** in Ursa Major.

August

August's Events

The summer is beginning to wind down, so it's best to soak up as much of the night sky as you can before the warm season ends. In fact, in the early morning sky, Orion — the most recognizable winter constellation — is already beginning to rise just before dawn.

Still, there's plenty of summer sky to enjoy.

On the morning of August 14, there is a beautiful conjunction of Mars and Jupiter, where the pair will be less than a quarter degree apart. You may also want to watch them in the few days before and after as they get closer and then begin to swing away from one another.

And of course, August is the month with one of the most anticipated meteor showers of the year, the Perseids, which peaks on the night of August 11 to 12. This shower can produce upward of 100 meteors an hour at its peak.

The warm temperatures make it an optimal time for viewers to sit out under the stars and enjoy the view.

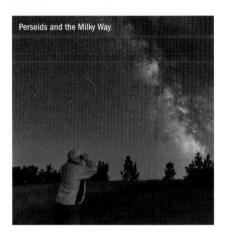
Perseids and the Milky Way

Calendar of Events

Day	Time (UTC)	Event
04	11:13	New Moon
05	22:04	Venus 1.9°S of Moon
07	17:23	Venus 5.7°N of Mercury
09	1:32	Moon at apogee: 405,300 km (251,842 mi.)
12		Perseid meteor shower
12	15:19	First quarter of the Moon
14	14:45	Jupiter 0.3°S of Mars
19	18:26	Full Moon
21	2:54	Saturn 0.4°S of Moon
21	5:05	Moon at perigee: 360,200 km (223,818 mi.)
26	9:26	Last quarter of the Moon

The Moon This Month

SUN	MON	TUES	WED	THURS	FRI	SAT
				1	2	3
4 New Moon	5	6	7	8	9	10
11	12 1st Quarter	13	14	15	16	17
18	19 Full Moon	20	21	22	23	24
25	26 Last Quarter	27	28	29	30	31

On the night of August 20 to 21, there will be a conjunction of the Moon and Saturn, where the pair will be very close together.

The new Moon falls on August 4, with the first quarter on August 12. The full Moon occurs on August 19, with the last quarter on August 26.

The Moon is at apogee on August 9 and at perigee on August 21.

A view of the Moon during the day

Helix Nebula

Highlights in the Southern Sky

Scorpius (the Scorpion) is beginning to sink toward the southwest, but **Sagittarius (the Archer)** is still fairly well placed. **Ptolemy's Cluster (Messier 7)** is low in the sky, but you can still see the **Sagittarius Star Cloud (Messier 24)** with the naked eye (if you're able to get away from the city) – or use binoculars, if you happen to have a pair. **Messier 25,** an open cluster, is another beautiful sight in Sagittarius. You can also find a fantastic globular cluster, **Messier 22,** just south of Messier 25.

Ophiuchus (the Serpent Bearer) is high in the sky and is where you can find the open clusters **Messier 10** and the **Summer Beehive Cluster (IC 4665)**.

In the nearby constellation **Scutum (the Shield)**, you can find the **Wild Duck Cluster (Messier 11)**, a big open cluster that is another a great target for binoculars.

Hercules is very well placed, with its familiar keystone shape. Grab a pair of binoculars to find the amazing globular cluster **Messier 13.** Between his knees is another great globular cluster, **Messier 92.**

Boötes (the Herdsman) is now in the west, next to **Corona Borealis (the Northern Crown)**.

Rising in the east, you can find **Capricornus (the Goat)** along with **Aquarius (the Water Bearer)**. An interesting target near Aquarius that can be seen through binoculars at dark-sky locations is the **Helix Nebula (NGC 7293)**, sometimes referred to as the "Eye of God" due to its appearance. This is a striking planetary nebula that formed after a star similar to our own Sun shed off its outer shell of gas, leaving behind a small white-dwarf star. View the faint object in late evening by first finding bright **Fomalhaut** in the constellation **Piscis Austrinus (the Southern Fish)**, which will be low on the southern horizon, then looking north 10 degrees.

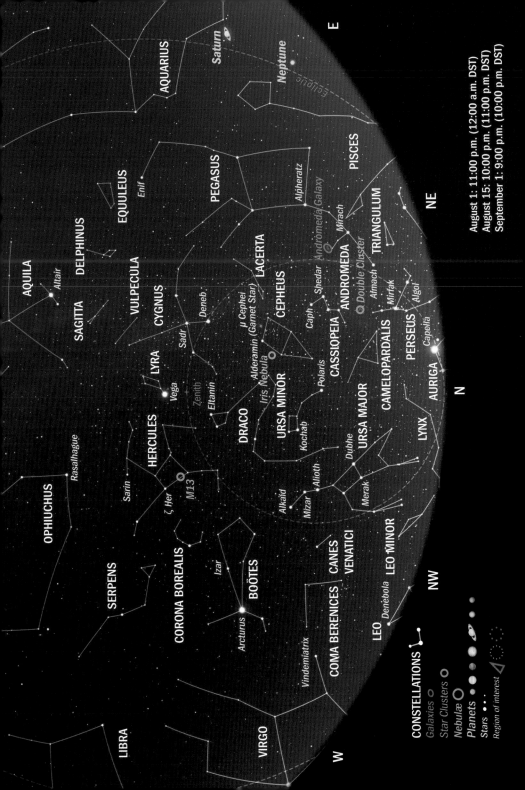

Iris Nebula

Highlights in the Northern Sky

Perseus (the Hero) is now rising higher in the northeast, along with the **Double Cluster (NGC 869 and NGC 884)**, a favorite of many amateur astronomers. The two clusters are roughly 7,000 light-years away and are easily visible with the unaided eye under even moderately dark-sky conditions.

Cassiopeia (the Queen) has now risen higher in the northeast and is easily identifiable with its distorted W shape. The constellations **Andromeda (the Princess)** and **Pegasus (the Flying Horse)** are also in prime locations in the east.

Using Cassiopeia as a guide, you can find the closest spiral galaxy to our Milky Way, the **Andromeda Galaxy (Messier 31)**, which is actually located in Andromeda (hence its name). **Shedar's** half of Cassiopeia points toward Messier 31. Or find **Mirach** in Andromeda, and look a palm's width higher. If you're under a dark sky, you can see the galaxy without the aid of binoculars; it will appear as a faint fuzzy spot. But even under city lights, binoculars will reveal the faint, but large, fuzzy oval. It's six full Moon diameters wide.

Cepheus (the King) is now fairly high in the sky, making it an ideal time to catch the "Garnet Star" **(Mu [μ] Cephei)** and, for astrophotographers, the **Iris Nebula (NGC 7023)**.

September

September's Events

The nights are now beginning to get noticeably longer — good news for stargazers.

Fall is a good time to enjoy the night sky, as the cooler nights allow for better viewing (since the heat and humidity of summer can make seeing fainter stars more challenging).

The summer constellations are beginning to set in the west, while the winter constellations are beginning to rise in the east in the early mornings.

Mercury is now visible just before sunrise, sitting low on the eastern horizon at the beginning of the month, though it will get lost in the Sun's glare by mid-month. However, Mars and Jupiter can be found high in the southeast in the hours before dawn.

In the evening sky, Saturn can be found hanging out in Aquarius, while Venus becomes lost in the glare of the sunset.

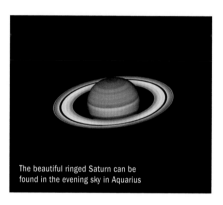

The beautiful ringed Saturn can be found in the evening sky in Aquarius

Calendar of Events

Day	Time (UTC)	Event
03	1:55	New Moon
05	01:59	Mercury 18.1°W of Sun (greatest western elongation)
05	10:13	Venus 1.3°N of Moon
05	14:55	Moon at apogee: 406,200 km (252,401 mi.)
08	04:00	Saturn at opposition
11	6:06	First quarter of the Moon
17	10:14	Saturn 0.3°S of Moon
18	2:34	Full Moon
18	2:45	Partial lunar eclipse
18	13:26	Moon at perigee: 357,300 km (222,016 mi.)
22	12:44	Autumnal (fall) equinox
24	18:50	Last quarter of the Moon

The Moon This Month

SUN	MON	TUES	WED	THURS	FRI	SAT
1	2	3 New Moon	4	5	6	7
8	9	10	11 1st Quarter	12	13	14
15	16	17	18 Full Moon	19	20	21
22	23	24 Last Quarter	25	26	27	28
29	30					

There is a partial lunar eclipse on September 18, which will be visible in the evening across North America. It will be difficult to notice much of a difference in the Moon's color brightness because only a small fraction of the Moon will enter the darkest part of Earth's shadow and most of this eclipse will occur as a penumbral lunar eclipse.

On September 8, Saturn will reach opposition, when it will shine brightest and look largest in a telescope for 2024. On September 17, the full Moon will shine very close to Saturn, and observers in western North America can enjoy the spectacle of the Moon occulting Saturn. On September 25, the Moon will be roughly 5 degrees from Mars in the morning sky.

The new Moon falls on September 3, with the first quarter on September 11. The full Moon is on September, with the last quarter on September 24.

The Moon is at apogee on September 5 and at perigee on September 18.

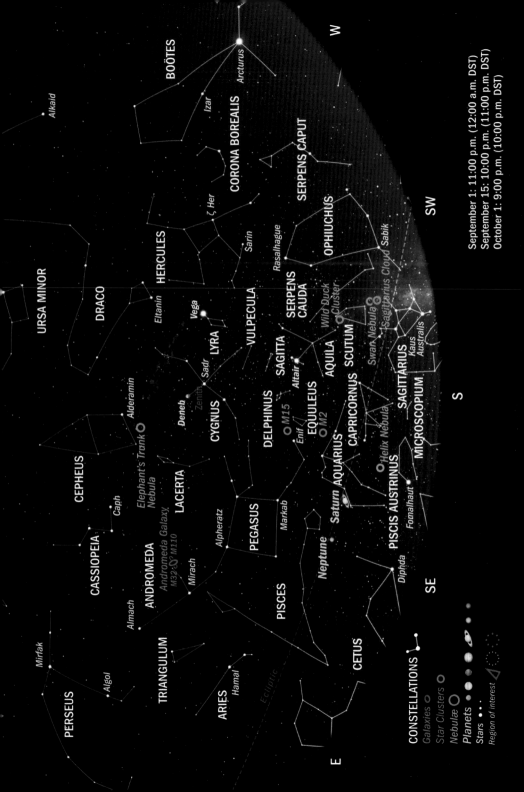

September 1: 11:00 p.m. (12:00 a.m. DST)
September 15: 10:00 p.m. (11:00 p.m. DST)
October 1: 9:00 p.m. (10:00 p.m. DST)

W

SW

S

SE

E

BOÖTES

Arcturus

Izar

Alkaid

CORONA BOREALIS

ζ Her

Sarin

SERPENS CAPUT

OPHIUCHUS

Rasalhague

Sabik

HERCULES

Eltanin

DRACO

URSA MINOR

Vega

LYRA

Sadr

Zenith

CYGNUS

Deneb

Alderamin

Elephant's Trunk
Nebula

CEPHEUS

Caph

LACERTA

Alpheratz

Mirach

ANDROMEDA

Andromeda Galaxy.
M32 · M110

CASSIOPEIA

Almach

Mirfak

Algol

PERSEUS

TRIANGULUM

Hamal

ARIES

VULPECULA

SAGITTA

Altair

AQUILA

SCUTUM

Wild Duck
Cluster

Swan Nebula

Sagittarius Cloud

SERPENS
CAUDA

Kaus
Australis

SAGITTARIUS

MICROSCOPIUM

DELPHINUS

M15

Enif

EQUULEUS

M2

CAPRICORNUS

Helix Nebula

AQUARIUS

Saturn

Neptune

Markab

PEGASUS

PISCES

PISCIS AUSTRINUS

Fomalhaut

Diphda

CETUS

Ecliptic

CONSTELLATIONS

Galaxies ○
Star Clusters ○
Nebulæ ○
Planets ●
Stars · • ●
Region of interest △

Highlights in the Southern Sky

The faint stars of **Aquarius (the Water Bearer)** and **Capricornus (the Sea Goat)** are now dominating the southern sky. One interesting target is **Messier 2**, one of the largest-known globular clusters, which spans roughly 175 light-years. It is also one of the oldest, believed to be about 13 billion years old. (Remember: The Universe is roughly 13.8 billion years old.)

Piscis Austrinus (the Southern Fish) can be found low on the horizon. It's more easily seen from the Southern Hemisphere, but northerners can see it, too, at this time of year. While it's not a bright constellation, it contains a bright star, **Fomalhaut**.

To the east is **Pisces (the Fishes)**, with **Pegasus (the Flying Horse)** sitting directly north along with **Andromeda (the Princess)**. You can find **Messier 15**, a globular cluster, in Pegasus and the remarkable **Andromeda Galaxy (Messier 31)** in Andromeda. If you happen to look at the Andromeda Galaxy through a telescope, you'll notice two smaller galaxies alongside it, **Messier 110** and **Messier 32.**

Sagittarius (the Archer) is sinking low in the southwest, but binoculars will still reveal the **Wild Duck Cluster (Messier 11)**, the **Sagittarius Star Cloud (Messier 24)** and the **Swan Nebula (Messier 17)**, also known as the **Omega Nebula.**

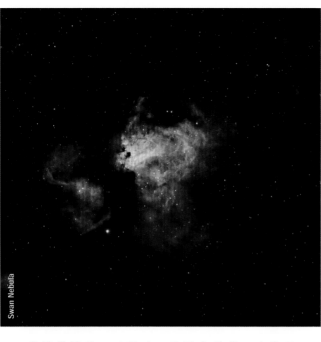

Swan Nebula

Highlights in the Northern Sky

You can find **Perseus (the Hero)** in the northeast now, with the **Double Cluster (NGC 869 and NGC 884)** well placed. These two open clusters can be seen with the unaided eye (best at dark-sky locations), but they make spectacular binocular targets.

Below Perseus, you may have noticed a particularly bright star. That would be **Capella**, the brightest star in **Auriga (the Charioteer)**, which is just beginning to rise in the northeast.

Nearby is **Cassiopeia (the Queen)**, where binoculars will reveal several open clusters, including **Messier 52** and **Messier 103**.

Directly north is **Cepheus (the King)**, which now looks like an upside-down house. This is a great time to look for **Mu (μ) Cephei**, or the "Garnet Star."

Turning to the northwest, you can find **Ursa Major (the Great Bear)** beginning to hug the horizon, with winding **Draco (the Dragon)** above it. Within Draco lies the **Cat's Eye Nebula (NGC 6543)**, a planetary nebula. The constellation **Hercules** is close by, where you can find two beautiful globular clusters, **Messier 13** and **Messier 92**.

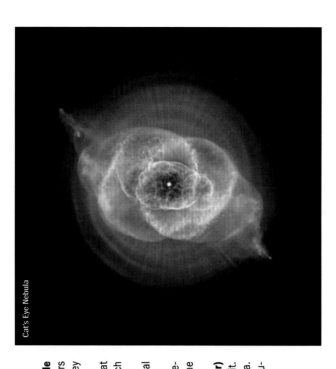

Cat's Eye Nebula

October

October's Events

It's time to bundle up and head outside to enjoy both the end of the summer constellations and the arrival of the winter constellations.

The Summer Triangle — with its three bright stars Deneb, Vega and Altair — is now in the southwest. Also in the southern sky, front and center, is Aquarius hosting Saturn. Ursa Major is low in the north, and Auriga with its brightest star Capella is rising in the northeast, along with Taurus — where you can find Jupiter in late evening. By the end of the month, Orion sits low in the east.

There's an annular solar eclipse on October 2, but unfortunately it won't be visible in North America (see page 35).

However, North American sky watchers can enjoy the annual meteor shower, the Orionids. Though not as spectacular as the summer's Perseids — or the coming Geminids — this shower can produce about 10 to 20 meteors an hour on its peak night, October 20 to 21.

Calendar of Events

Day	Time (UTC)	Event
02	18:46	Annular solar eclipse (not observable from North America)
02	18:49	New Moon
02	19:40	Moon at apogee: 406,500 km (252,587 mi.)
05	20:26	Venus 3.3°N of Moon
10	18:55	First quarter of the Moon
14	18:05	Saturn 0.1°S of Moon
17	0:46	Moon at perigee: 357,200 km (221,954 mi.)
17	11:26	Full Moon
21		Orionid meteor shower
23	19:55	Mars 4.2°S of Moon
24	8:03	Last quarter of the Moon
29	22:50	Moon at apogee: 406,200 km (252,401 mi.)

The Moon This Month

SUN	MON	TUES	WED	THURS	FRI	SAT
		1	2 New Moon	3	4	5
6	7	8	9	10 1st Quarter	11	12
13	14	15	16	17 Full Moon	18	19
20	21	22	23	24 Last Quarter	25	26
27	28	29	30	31		

On October 14, you can find an almost-full Moon and Saturn about 4 degrees apart in Aquarius. And on October 23 and 24, Mars and a waning gibbous Moon meet up in the early morning.

The new Moon falls on October 2, with the first quarter occurring on October 10. The full Moon is on October 17, and the last quarter occurs on October 24.

The Moon is at apogee twice this month: October 2 and 29. It is at perigee on October 17.

A size comparison of a full Moon at apogee (farthest from Earth) and the full Moon at perigee (closest to Earth). The full Moon will be at perigee on October 17

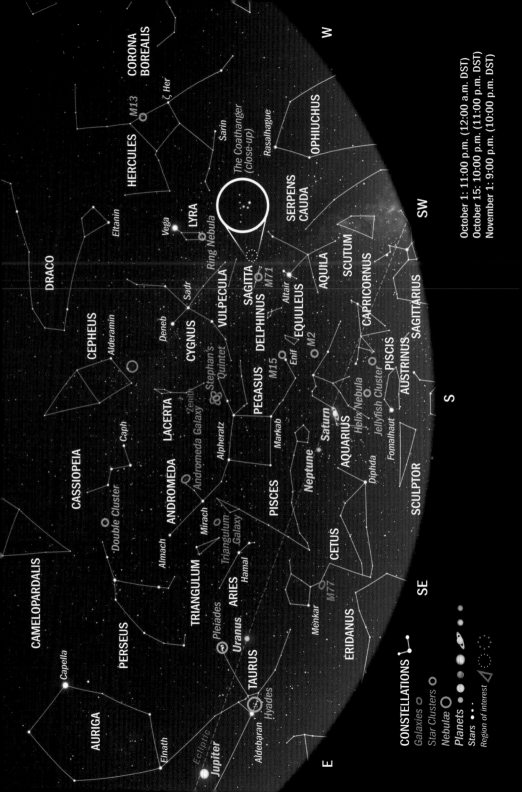

CONSTELLATIONS
Galaxies ⊙
Star Clusters ○
Nebulae ○
Planets ● ● ● ●
Stars ● • ·
Region of interest

W

SW

S

SE

E

October 1: 11:00 p.m. (12:00 a.m. DST)
October 15: 10:00 p.m. (11:00 p.m. DST)
November 1: 9:00 p.m. (10:00 p.m. DST)

CORONA BOREALIS
HERCULES
M13
ζ Her
Sarin
Rasalhague
OPHIUCHUS
SERPENS CAUDA
The Coathanger (close-up)
Eltanin
DRACO
Vega
LYRA
Ring Nebula
AQUILA
SCUTUM
Sadr
Deneb
CYGNUS
VULPECULA
SAGITTA
M71
DELPHINUS
Altair
EQUULEUS
Enif
M2
CAPRICORNUS
SAGITTARIUS
CEPHEUS
Alderamin
Stephan's Quintet
Zenith
PEGASUS
M15
Markab
PISCIS AUSTRINUS
LACERTA
Andromeda Galaxy
Alpheratz
Saturn
Neptune
AQUARIUS
Helix Nebula
Jellyfish Cluster
Fomalhaut
CASSIOPEIA
Caph
Double Cluster
ANDROMEDA
Mirach
Triangulum Galaxy
PISCES
Diphda
SCULPTOR
Almach
TRIANGULUM
ARIES
Hamal
CETUS
Menkar
M77
ERIDANUS
CAMELOPARDALIS
PERSEUS
Pleiades
Uranus
Jupiter
TAURUS
Aldebaran
Hyades
Ecliptic
Capella
AURIGA
Elnath

Messier 77

Highlights in the Southern Sky

Pegasus (the Flying Horse) is high in the south, making it a great time to seek out **Messier 15** (also known as the **Pegasus Cluster**). This globular cluster lies some 33,600 light-years away from Earth and is one of the densest-known clusters, containing more than 30,000 stars. It may also host an intermediate-size black hole, over 4,000 times more massive than our Sun, in its center.

Not too far away is the tiny constellation **Sagitta (the Arrow)**, which sits within the **Summer Triangle** asterism. Though somewhat unremarkable, you can find **Messier 71**, one of the smallest-known globular clusters, within Sagitta.

If you have a pair of binoculars, you can look for an asterism called the **Coathanger** within **Vulpecula (the Little Fox)**. The binoculars will make it very clear how this asterism got its name.

To the southeast, you'll find **Cetus (the Whale)**, and within it is **Messier 77**, also known as **Cetus A**. This is one of the largest galaxies, sitting at a distance of 47 million light-years away. Binoculars will reveal it under dark-sky locations as a faint fuzz, while telescopes will show the structure of this barred galaxy.

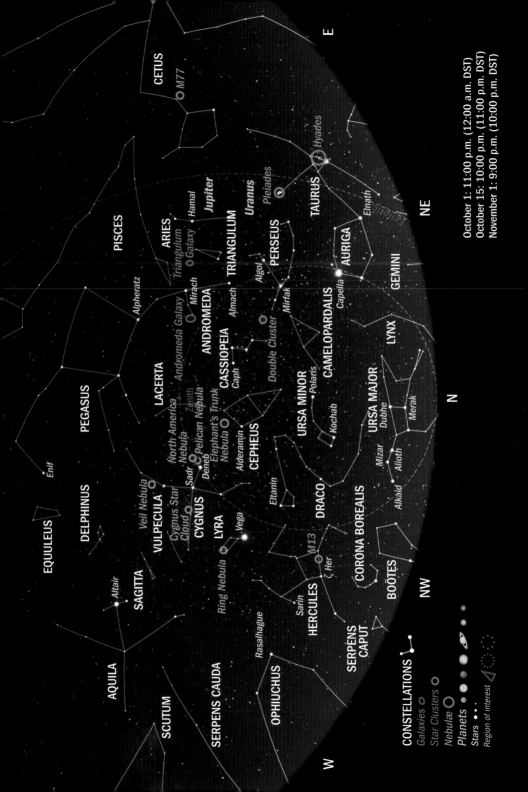

Highlights in the Northern Sky

It's a great time to turn your binoculars to **Cygnus (the Swan)**, which has a lot of celestial treats, including the **Veil Nebula (NGC 6992 and NGC 6960)**, the **Cygnus Star Cloud**, the **North America Nebula (NGC 7000)**, the **Pelican Nebula (IC 5070)** and several open clusters.

Nearby is **Lyra (the Lyre)**, where a telescope will reveal the **Ring Nebula (Messier 57)**, the result of a star with the mass of our Sun shedding off its layers and leaving behind a small white dwarf star surrounded by a shell of glowing gas.

Ursa Major (the Great Bear) and its asterism the **Big Dipper** are low in the north. **Auriga (the Charioteer)** is prominent in the northeast with its star **Capella** shining brightly.

Perseus (the Hero) is high in the east where you can find the **Double Cluster (NGC 869 and NGC 884)**. High above is the **Andromeda Galaxy (Messier 31)**. Nearby is the small constellation **Triangulum (the Triangle)**, where you can find a favorite target for many astrophotographers, the **Triangulum Galaxy (Messier 33)**.

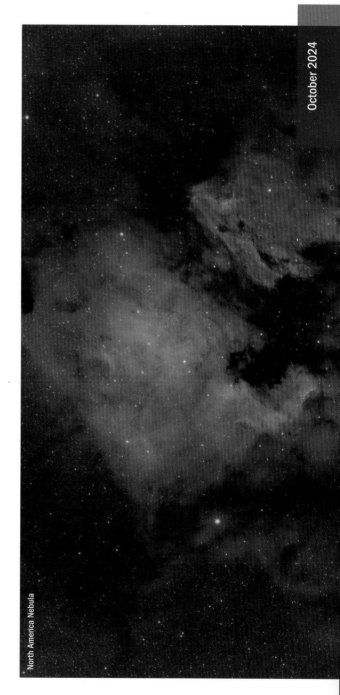

North America Nebula

October 2024

November

November's Events

The winter constellations are now here.

Taurus begins to rise shortly after sunset, where you'll also find Jupiter, and right behind it is Orion and Gemini. Mars is starting to rise in the east later in the night, and Saturn continues to dominate Aquarius.

Cygnus is in the west, but by the end of the month, it begins to disappear — confirmation that winter is truly upon us.

There are plenty of meteor showers to enjoy this month, so if you have clear skies, it's a good time to bundle up and search for "shooting stars." The Southern Taurids peak on the moonless night of November 4 to 5, and the Northern Taurids peak on the moonlit night of November 11 to 12. But wait, there's more: The Leonids peak on the brightly moonlit night of November 16 to 17.

Taurus

Calendar of Events

Day	Time (UTC)	Event
01	12:47	New Moon
05	0:16	Venus 3.4°N of Moon
05		Southern Taurid meteor shower
09	5:56	First quarter of the Moon
11	1:36	Saturn 0.1°N of Moon
12		Northern Taurid meteor shower
14	11:18	Moon at perigee: 360,100 km (223,756 mi.)
15	21:29	Full Moon
16	07:59	Mercury 22.5°E of Sun (greatest eastern elongation)
17		Leonid meteor shower
20	21:07	Mars 2.6°S of Moon
23	1:28	Last quarter of the Moon
26	11:56	Moon at apogee: 405,300 km (251,842 mi.)

The Moon This Month

SUN	MON	TUES	WED	THURS	FRI	SAT
					1 New Moon	2
3	4	5	6	7	8	9 1st Quarter
10	11	12	13	14	15 Full Moon	16
17	18	19	20	21	22	23 Last Quarter
24	25	26	27	28	29	30

There is a beautiful conjunction on November 11, where you will find Saturn and the roughly 10-day-old Moon so close together, they almost seem to touch in the sky. Observers from Florida can see the Moon occult Saturn.

On November 20, you will find Mars and the waning gibbous Moon separated by roughly 5 degrees.

The month starts off with a new Moon, followed by the first quarter on November 9. The full Moon brightens up the sky on November 15, with the last quarter falling on November 23.

The Moon is at perigee on November 14 and at apogee on November 26.

The 2025 edition of *Night Sky Almanac* is in stores this month. Be sure to grab your copy, wherever books are sold.

A stunning halo surrounding the Moon

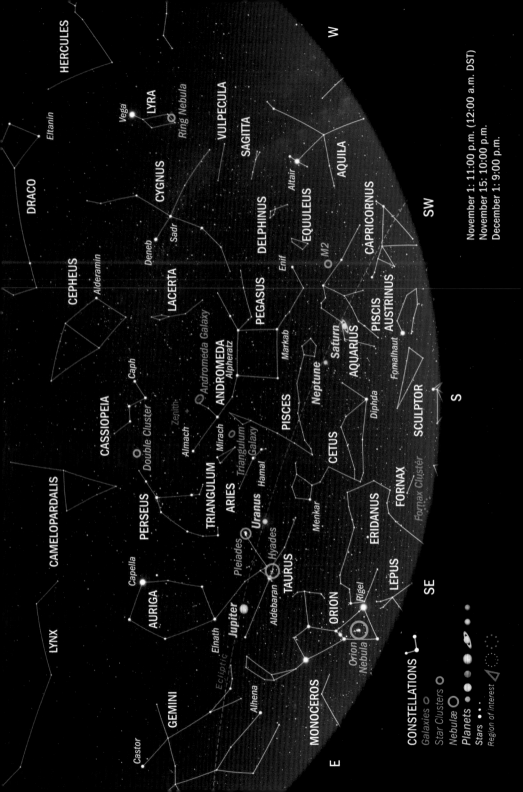

Fornax Cluster

Highlights in the Southern Sky

Due south in late evening, you can find **Eridanus (the River Eridanus)** winding its way through the sky. The constellation begins near the star **Rigel**, the right "foot" of **Orion (the Hunter)**, and jogs toward **Cetus (the Whale)** and then down to the horizon.

Eridanus skirts tiny **Fornax (the Furnace)**, a constellation that was added in the 18th century. While somewhat unremarkable, it is home to the **Fornax Cluster**, which contains 58 galaxies and is one of the closest clusters to Earth at approximately 62 million light-years away.

November is once again a great time to spot the popular star clusters the **Hyades (Melotte 25)** and the **Pleiades (Messier 45)**. A pair of binoculars will reveal the stunning nearby open cluster the Hyades, marked by **Aldebaran**, the brightest star in **Taurus (the Bull)**, though the star itself is not part of the cluster.

The Pleiades, also known as the "Seven Sisters," is another fantastic target for binoculars, though it can be seen easily without them, even from light-polluted cities. Photographs reveal a stunning display of blue wispy gas and dust around the stars.

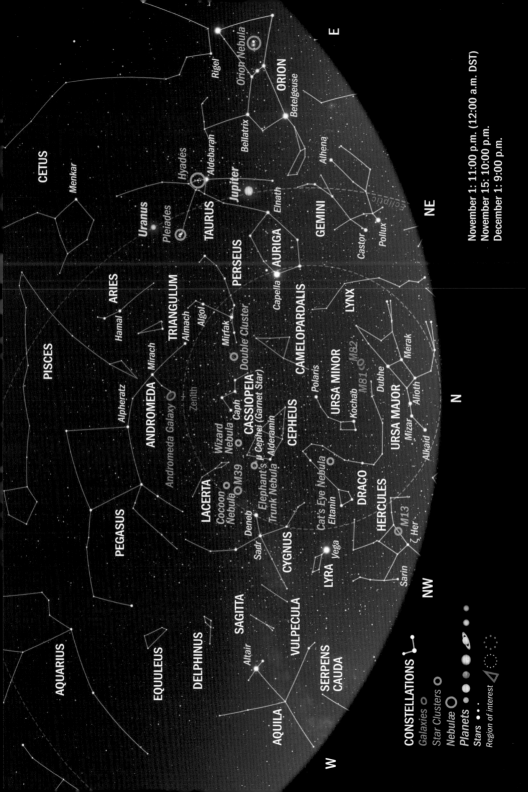

Cocoon Nebula

Highlights in the Northern Sky

Ursa Major (the Great Bear) is beginning to creep back up in the north-east. Late night would be an opportune time to seek out two spectac-ular galaxies, **Messier 81 (the Cigar Galaxy)** and **Messier 82 (Bode's Galaxy)**. It's also a good time to find the **Cat's Eye Nebula (NGC 6543)** in **Draco (the Dragon)**.

The constellation **Cepheus (the King)** can now be found high in the northwest, with nearby **Cygnus (the Swan)** beginning to sink toward the horizon. But it's a good time to find **Messier 39**, a bright open cluster in Cygnus. The large cluster is roughly 824 light-years from Earth and makes a great binocular target. Also within Cygnus is the **Cocoon Nebula (IC 5146)**, which lies roughly 3,300 light-years from Earth. It can be seen through small telescopes and is a favorite astrophotography target.

Within Cepheus, you can find the "Garnet Star" **(Mu [μ] Cephei)** and beside it, the **Elephant's Trunk Nebula**. Not too far away is the **Wizard Nebula (NGC 7380)**. This nebula is home to an open cluster believed to be only 4 million years old.

December

December's Events

Orion, the mighty hunter of the winter sky, is now fully above the horizon, joined by Taurus and its stunning open clusters the Hyades and the Pleiades.

When it comes to planets, you can see four of the brightest planets this month. Venus is an "evening star" in the west, before setting early in the evening. Saturn is sitting pretty in Aquarius, while Jupiter is hanging out in Taurus toward the east. And finally, Mars rises in Cancer later in the evening.

On December 21, Mercury will reach its highest point in the southeastern early morning sky, though it will be close to the horizon, making it tricky to see. Over the coming weeks, the Sun's closest companion will still be visible while it travels lower and lower along the horizon.

December is the time to enjoy one of the best meteor showers of the year: the Geminids. This shower rivals the much more enjoyable (weather-wise) Perseids, producing some 150 meteors per hour at its peak. The Geminids is one of the most reliable showers and can produce stunning fireballs. The downside is the Moon will be almost full on the peak night.

Winter solstice is on December 21.

Calendar of Events

Day	Time (UTC)	Event
01	6:21	New Moon
04	22:40	Venus 2.4°N of Moon
07	20:00	Jupiter at opposition
08	8:49	Saturn 0.3°S of Moon
08	15:27	First quarter of the Moon
12	13:18	Moon at perigee: 365,400 km (227,049 mi.)
14		Geminid meteor shower
15	9:02	Full Moon
18	8:46	Mars 1°S of Moon
21	9:20	Winter solstice
22		Ursid meteor shower
22	22:18	Last quarter of the Moon
24	7:25	Moon at apogee: 404,500 km (251,345 mi.)
25	02:59	Mercury 22°W of Sun (greatest western elongation)
30	22:27	New Moon

The Moon This Month

SUN	MON	TUES	WED	THURS	FRI	SAT
1 New Moon	2	3	4	5	6	7
8 1st Quarter	9	10	11	12	13	14
15 Full Moon	16	17	18	19	20	21
22 Last Quarter	23	24	25	26	27	28
29	30 New Moon	31				

If you're looking for Mercury, you could try to find it on December 28, when the Moon will be roughly 7 degrees to Mercury's right. They'll be extremely low on the horizon, so you'll need an unobstructed view. On December 7 Jupiter will reach opposition, when it will shine brightest and look largest in a telescope for 2024.

The month begins with a new Moon. Then on December 8, the Moon will be at first quarter, with the last full Moon of 2024 falling on December 15. Finally, the last quarter is on December 22 and the new Moon is on December 30.

The Moon will be at perigee on December 12 and at apogee on December 24.

A close-up of the waxing gibbous Moon

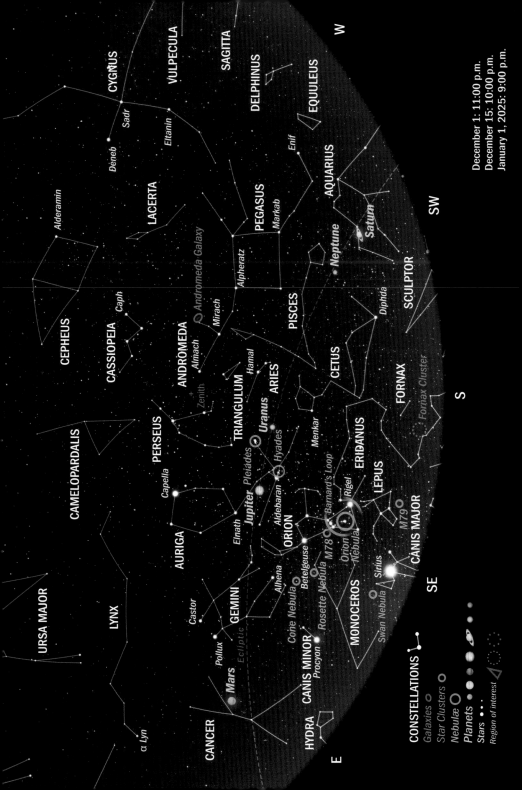

Barnard's Loop and Messier 78

Highlights in the Southern Sky

We end the year as we began it, with **Orion (the Hunter)**. The prominent winter constellation is rising in the southeast and will become the main object in the south by the end of the month.

Orion is home to a wealth of targets, most notably the star-forming **Orion Nebula (Messier 42)**. You can also find **Messier 78** near his belt. It's another gorgeous nebula that is best seen through telescopes. There is also the giant **Barnard's Loop** which makes a stunning target for astrophotographers.

Monoceros (the Unicorn) can be found to the east of Orion. The constellation is home to many beautiful targets including the faint **Seagull Nebula (IC 2177)** and the brighter **Rosette Nebula (NGC 2237)** and **Cone Nebula (NGC 2264)**.

Found south of Orion, in the southeastern sky, is **Lepus (the Hare)**. Within it is **Messier 79**, a globular cluster that contains roughly 150,000 stars. It can be seen as a small fuzzy patch through binoculars. A medium-sized telescope will reveal many of the stars.

You've likely noticed **Sirius** rising in the southeast. It is the brightest star in **Canis Major (the Great Dog)** and is overall the brightest star in the night sky.

To the southwest, you can find the faint stars of **Cetus (the Whale)** and **Pisces (the Fish)**.

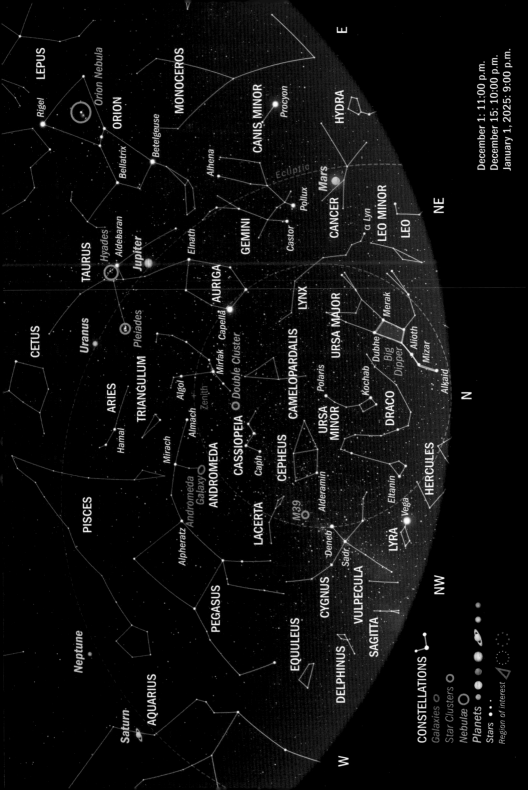

December 1: 11:00 p.m.
December 15: 10:00 p.m.
January 1, 2025: 9:00 p.m.

Highlights in the Northern Sky

Cassiopeia (the Queen) is now high in the sky looking like an "M," on top of **Cepheus (the King)**, while **Draco (the Dragon)** is low in the north stretching from the horizon up to **Ursa Minor (the Little Bear)** and then toward **Ursa Major (the Great Bear)**. Ursa Major's **Big Dipper** shines below **Lynx (the Lynx)** and, below that is **Leo Minor (the Little Lion)**. **Perseus (the Hero)** is almost at the zenith, which makes it a great

time to grab a pair of binoculars and find the **Double Cluster (NGC 869 and NGC 884)**. If you're at a dark-sky location, you'll have no problem seeing it without binoculars.

Cygnus (the Swan) has almost disappeared below the northwestern horizon, but you may still view the bright open cluster **Messier 39** above **Deneb**.

Double Cluster

December 2024

The Messier Catalog

Name	Traditional Name	Type	Constellation
Messier 1 (NGC 1952)	Crab Nebula	Supernova remnant	Taurus
Messier 2 (NGC 7089)		Globular cluster	Aquarius
Messier 3 (NGC 5272)		Globular cluster	Canes Venatici
Messier 4 (NGC 6121)		Globular cluster	Scorpius
Messier 5 (NGC 5904)		Globular cluster	Serpens
Messier 6 (NGC 6405)	Butterfly Cluster	Open cluster	Scorpius
Messier 7 (NGC 6475)	Ptolemy Cluster	Open cluster	Scorpius
Messier 8 (NGC 6523)	Lagoon Nebula	Emission nebula with cluster	Sagittarius
Messier 9 (NGC 6333)		Globular cluster	Ophiuchus
Messier 10 (NGC 6254)		Globular cluster	Ophiuchus
Messier 11 (NGC 6705)	Wild Duck Cluster	Open cluster	Scutum
Messier 12 (NGC 6218)	Gumball Globular	Globular cluster	Ophiuchus
Messier 13 (NGC 6205)	Hercules	Globular cluster	Hercules
Messier 14 (NGC 6402)		Globular cluster	Ophiuchus
Messier 15 (NGC 7078)	Great Pegasus Cluster	Globular cluster	Pegasus
Messier 16 (NGC 6611)	Eagle Nebula	Emission nebula with cluster	Serpens
Messier 17 (NGC 6618)	Omega Nebula	Emission nebula with cluster	Sagittarius
Messier 18 (NGC 6613)		Open cluster	Sagittarius
Messier 19 (NGC 6273)		Globular cluster	Ophiuchus
Messier 20 (NGC 6514)	Trifid Nebula	Emission, reflection and dark nebula with cluster	Sagittarius
Messier 21 (NGC 6531)		Open cluster	Sagittarius
Messier 22 (NGC 6656)	Sagittarius Cluster	Globular cluster	Sagittarius
Messier 23 (NGC 6494)		Open cluster	Sagittarius
Messier 24 (IC 4715)	Sagittarius Star Cloud	Milky Way star cloud	Sagittarius
Messier 25 (IC 4725)		Open cluster	Sagittarius
Messier 26 (NGC 6694)		Open cluster	Scutum
Messier 27 (NGC 6853)	Dumbbell Nebula	Planetary nebula	Vulpecula
Messier 28 (NGC 6626)		Globular cluster	Sagittarius
Messier 29 (NGC 6913)		Open cluster	Cygnus

Name	Traditional Name	Type	Constellation
Messier 30 (NGC 7099)		Globular cluster	Capricornus
Messier 31 (NGC 224)	Andromeda Galaxy	Spiral galaxy	Andromeda
Messier 32 (NGC 221)	Le Gentil	Dwarf elliptical galaxy	Andromeda
Messier 33 (NGC 598)	Triangulum Galaxy	Spiral galaxy	Triangulum
Messier 34 (NGC 1039)		Open cluster	Perseus
Messier 35 (NGC 2168)		Open cluster	Gemini
Messier 36 (NGC 1960)	Pinwheel Cluster	Open cluster	Auriga
Messier 37 (NGC 2099)		Open cluster	Auriga
Messier 38 (NGC 1912)	Starfish Cluster	Open cluster	Auriga
Messier 39 (NGC 7092)		Open cluster	Cygnus
Messier 40	Winnecke 4	Double star	Ursa Major
Messier 41 (NGC 2287)		Open cluster	Canis Major
Messier 42 (NGC 1976)	Orion Nebula	Emission-reflection nebula	Orion
Messier 43 (NGC 1982)	De Mairan's Nebula	Emission-reflection nebula	Orion
Messier 44 (NGC 2632)	Beehive Cluster	Open cluster	Cancer
Messier 45	Pleiades	Open cluster	Taurus
Messier 46 (NGC 2437)		Open cluster	Puppis
Messier 47 (NGC 2422)		Open cluster	Puppis
Messier 48 (NGC 2548)		Open cluster	Hydra
Messier 49 (NGC 4472)		Elliptical galaxy	Virgo
Messier 50 (NGC 2323)	Heart-Shaped Cluster	Open cluster	Monoceros
Messier 51 (NGC 5194, NGC 5195)	Whirlpool Galaxy	Spiral galaxy	Canes Venatici
Messier 52 (NGC 7654)		Open cluster	Cassiopeia
Messier 53 (NGC 5024)		Globular cluster	Coma Berenices
Messier 54 (NGC 6715)		Globular cluster	Sagittarius
Messier 55 (NGC 6809)	Summer Rose Star	Globular cluster	Sagittarius
Messier 56 (NGC 6779)		Globular cluster	Lyra
Messier 57 (NGC 6720)	Ring Nebula	Planetary nebula	Lyra
Messier 58 (NGC 4579)		Barred spiral galaxy	Virgo
Messier 59 (NGC 4621)		Elliptical galaxy	Virgo

Name	Traditional Name	Type	Constellation
Messier 60 (NGC 4649)		Elliptical galaxy	Virgo
Messier 61 (NGC 4303)		Spiral galaxy	Virgo
Messier 62 (NGC 6266)		Globular cluster	Ophiuchus
Messier 63 (NGC 5055)	Sunflower Galaxy	Spiral galaxy	Canes Venatici
Messier 64 (NGC 4826)	Black Eye Galaxy	Spiral galaxy	Coma Berenices
Messier 65 (NGC 3623)		Barred spiral galaxy	Leo
Messier 66 (NGC 3627)		Barred spiral galaxy	Leo
Messier 67 (NGC 2682)	King Cobra Cluster	Open cluster	Cancer
Messier 68 (NGC 4590)		Globular cluster	Hydra
Messier 69 (NGC 6637)		Globular cluster	Sagittarius
Messier 70 (NGC 6681)		Globular cluster	Sagittarius
Messier 71 (NGC 6838)		Globular cluster	Sagittarius
Messier 72 (NGC 6981)		Globular cluster	Aquarius
Messier 73 (NGC 6994)		Asterism	Aquarius
Messier 74 (NGC 628)	Phantom Galaxy	Spiral galaxy	Pisces
Messier 75 (NGC 6864)		Globular cluster	Sagittarius
Messier 76 (NGC 650, NGC 651)	Little Dumbbell Nebula	Planetary nebula	Perseus
Messier 77 (NGC 1068)	Cetus A	Spiral galaxy	Cetus
Messier 78 (NGC 2068)		Reflection nebula	Orion
Messier 79 (NGC 1904)		Globular cluster	Lepus
Messier 80 (NGC 6093)		Globular cluster	Scorpius
Messier 81 (NGC 3031)	Bode's Galaxy	Spiral galaxy	Ursa Major
Messier 82 (NGC 3034)	Cigar Galaxy	Starburst irregular galaxy	Ursa Major
Messier 83 (NGC 5236)	Southern Pinwheel Galaxy	Barred spiral galaxy	Hydra
Messier 84 (NGC 4374)		Lenticular or elliptical galaxy	Virgo
Messier 85 (NGC 4382)		Lenticular or elliptical galaxy	Coma Berenices
Messier 86 (NGC 4406)		Lenticular or elliptical galaxy	Virgo
Messier 87 (NGC 4486)	Virgo A	Elliptical galaxy	Virgo

Name	Traditional Name	Type	Constellation
Messier 88 (NGC 4501)		Spiral galaxy	Coma Berenices
Messier 89 (NGC 4552)		Elliptical galaxy	Virgo
Messier 90 (NGC 4569)		Spiral galaxy	Virgo
Messier 91 (NGC 4548)		Barred spiral galaxy	Coma Berenices
Messier 92 (NGC 6341)		Globular cluster	Hercules
Messier 93 (NGC 2447)		Open cluster	Puppis
Messier 94 (NGC 4736)	Cat's Eye Galaxy	Spiral galaxy	Canes Venatici
Messier 95 (NGC 3351)		Barred spiral galaxy	Leo
Messier 96 (NGC 3368)		Spiral galaxy	Leo
Messier 97 (NGC 3587)	Owl Nebula	Planetary nebula	Ursa Major
Messier 98 (NGC 4192)		Spiral galaxy	Coma Berenices
Messier 99 (NGC 4254)	Coma Pinwheel	Spiral galaxy	Coma Berenices
Messier 100 (NGC 4321)		Spiral galaxy	Coma Berenices
Messier 101 (NGC 5457)	Pinwheel Galaxy	Spiral galaxy	Ursa Major
Messier 102 (NGC 5866)	Spindle Galaxy	Lenticular galaxy	Draco
Messier 103 (NGC 581)		Open cluster	Cassiopeia
Messier 104 (NGC 4594)	Sombrero Galaxy	Spiral galaxy	Virgo
Messier 105 (NGC 3379)		Elliptical galaxy	Leo
Messier 106 (NGC 4258)		Spiral galaxy	Canes Venatici
Messier 107 (NGC 6171)		Globular cluster	Ophiuchus
Messier 108 (NGC 3556)	Surfboard Galaxy	Barred spiral galaxy	Ursa Major
Messier 109 (NGC 3992)		Barred spiral galaxy	Ursa Major
Messier 110 (NGC 205)	Edward Young Star	Dwarf elliptical galaxy	Andromeda

Glossary

annular solar eclipse: A kind of partial solar eclipse during which the Moon does not completely cover the Sun's disk but leaves a ring of sunlight around the Moon.

aphelion: For an object orbiting the Sun, the point in its orbit that is farthest from the Sun.

apogee: When the Moon (or any satellite of Earth) is at its most distant in its monthly orbit around Earth.

arcsecond (or second of arc): A tiny angle equal to 1/3600 of a degree, used to measure the separation of double stars and the apparent diameters of Solar System objects.

asterism: A group of stars within a constellation (or sometimes from more than one constellation) that forms its own distinct pattern.

astronomical unit (au): A unit of distance that uses the average distance between Earth and the Sun as its base metric. One astronomical unit is approximately 150 million kilometers (93.2 million miles).

autumnal (fall) equinox: The date in September marking the end of summer and the beginning of autumn, when the Sun illuminates the Northern and Southern Hemispheres equally. The lengths of the day and night are equal, and it's one of two instances of the year that the Sun has a declination of 0 degrees.

binary star system: A double-star system in which two stars are gravitationally bound to each other and orbit a common center of mass. Many double-star systems have multiple components.

conjunction: Technically speaking a conjunction is when two objects have the same right ascension, but amateur astronomers also use the term when there is a close approach of two or more celestial objects.

constellation: A group of stars that make an imaginary image in the night sky. The International Astronomical Union (IAU) has designated the boundaries between 88 official constellations covering the celestial sphere.

degree: A unit of measurement used to measure the distance between objects or the position of objects in astronomy. The entire sky spans 360 degrees, and up to about 180 degrees of sky is visible from any given point on Earth with an unobstructed horizon.

double star (binary star): A pair of stars with small angular separation (from fractions of an arcsecond to hundreds of arcseconds). Optical doubles are chance alignments of stars with no physical connection to each other. True binary stars are gravitationally bound (see binary star system).

fireball: A very bright meteor, generally brighter than magnitude –4.0 (about the brightness of Venus).

galaxy: An enormous system of gas, dust and billions of stars and their planetary systems, all held together by gravity.

globular cluster: Old star systems at the edges of spiral galaxies that can contain anywhere from thousands to millions of stars, packed in a close, roughly spherical form and held together by gravity.

greatest eastern elongation: When an inner planet (Mercury or Venus) is farthest from the Sun in the evening sky.

greatest western elongation: When an inner planet (Mercury or Venus) is farthest from the Sun in the morning sky.

light-year: A unit of distance that uses how far light travels in one Earth year as its base metric. One light-year is about 9 trillion kilometers (6 trillion miles). Objects in outer space are so far apart that it takes a long time for their light to reach Earth, and so the farther an object is, the farther in the past that observers on Earth are seeing it. As an example, the Andromeda Galaxy (Messier 31) is 2.5 million light-years away, so when observers look at it in the sky, they are seeing it as it appeared 2.5 million years ago.

magnitude: The apparent brightness of an object in the sky. Magnitude is measured on a scale where the higher the number, the fainter the object appears. Objects with negative numbers are brighter than those with positive numbers.

meteor shower: An atmospheric event during which a number of meteors can be seen radiating from one point in the sky. Meteor showers occur when Earth, at a particular point in its orbit, crosses a stream of particles left over from a passing comet or asteroid.

meteor shower peak: The best time to view a meteor shower, when meteor activity will be at its highest.

nebula: A cloud of dust and gas visible in the night sky either as a bright patch or a dark shadow against other luminous matter. To learn more about the different types of nebulae, see page 42.

occultation: When one object appears to pass in front of another more distant object, obscuring the observer's view. For example, a lunar occultation is when the Moon appears to pass in front of a more distant object, like a planet or star.

open cluster: A star system that contains anything from a dozen to hundreds of stars, in which the stars are spread out. Open clusters are mostly found near the galactic plane, the plane on which most of a galaxy's mass lies.

opposition: When an outer planet (Mars, Jupiter, Saturn, Uranus or Neptune), a dwarf planet or a minor planet is opposite the Sun in the sky in right ascension.

partial lunar eclipse: A lunar eclipse during which the Earth's shadow partially covers the Moon.

partial solar eclipse: A solar eclipse during which the Moon only partially covers the Sun.

perigee: When the Moon (or any satellite of Earth) is at its closest in its monthly orbit around Earth.

perihelion: For an object orbiting the Sun, the point in its orbit that is nearest to the Sun.

star: A luminous ball of plasma held together by its own gravity and fueled initially by the nuclear fusion of hydrogen into helium at its core. Stars come in a vast variety of sizes, luminosities and temperatures, typically ranging from dwarf sizes (that are as little as 10 percent the mass of the Sun) to hypergiants (that can be 100 or more times the mass of the Sun). To learn more about the different types of stars, see pages 14–15.

summer solstice: The point during the year that a particular hemisphere is most tilted toward the Sun. This takes place in June in the Northern Hemisphere.

supernova: A powerful and luminous explosion that occurs during the last evolutionary stages of a massive star when its brightness increases immensely and it throws off most of its mass; a supernova can also be when a white dwarf is triggered into runaway nuclear fusion by the accretion of material from a binary companion; or by a stellar merger.

total lunar eclipse: A lunar eclipse during which the Earth's shadow completely covers the Moon. Such an eclipse can last for hours.

total solar eclipse: A solar eclipse during which the Moon covers the entire face of the Sun, revealing the Sun's corona and prominences. Solar eclipses can last from a few seconds to about seven minutes maximum at a specific location.

variable star: A star whose apparent magnitude varies over time. The variability might be caused by physical changes to the star itself or by how it rotates or interacts with nearby objects.

vernal (spring) equinox: The date in March marking the end of winter and the beginning of spring, when the Sun illuminates the Northern and Southern Hemispheres equally. The lengths of the day and night are equal, and it's one of two instances of the year that the Sun has a declination of 0 degrees

winter solstice: The point during the year that a particular hemisphere is most tilted away from the Sun. This takes place in December in the Northern Hemisphere.

Zenithal Hourly Rate (ZHR): The rate of meteors a shower would produce per hour under clear, dark skies and with the radiant at the zenith.

The Greek Alphabet (lowercase)

α	alpha	ζ	zeta	λ	lambda	π	pi	φ	phi
β	beta	η	eta	μ	mu	ρ	rho	χ	chi
γ	gamma	θ	theta	ν	nu	σ	sigma	ψ	psi
δ	delta	ι	iota	ξ	xi	τ	tau	ω	omega
ε	epsilon	κ	kappa	o	omicron	υ	upsilon		

Resources

American Astronomical Society — aas.org — An organization of professional astronomers seeking to enhance and share humanity's understanding of the Universe

American Meteor Society — amsmeteors.org — A non-profit scientific organization that informs, encourages and supports research in Meteor Astronomy

Eclipsewise.com — eclipsewise.com — Astronomer Fred Espenak's site dedicated to predictions and information on eclipses of the Sun and Moon

European Southern Observatory — eso.org — An intergovernmental research organization for ground-based astronomy that shares recent research, news and images

European Space Agency (ESA) — esa.int — The main site for the European Space Agency that shares news, research and launches

Hubblesite.org — hubblesite.org — NASA's main site for the Hubble Space Telescope, including news and images

James Webb Space Telescope — webb.nasa.gov — NASA's main site for the James Webb Space Telescope, including the latest images

NASA — nasa.gov — An update on all NASA missions, including astronomical events, news and launches

The Planetary Society — planetary.org — An international, non-profit organization that promotes the exploration of space through education, advocacy and research

The Royal Astronomical Society of Canada — rasc.ca — Home to Canada's national astronomical association

Spaceweather.com — spaceweather.com — An update on space weather, including solar flares, coronal mass ejections and noctilucent cloud forecasts

Space Weather Prediction Center — swpc.noaa.gov — The National Oceanic and Atmospheric Administration's site on forecasts for space weather, including potential geomagnetic storms

Solar and Heliospheric Observatory — soho.nascom.nasa.gov — A collaborative project between the ESA and NASA to study the Sun and its interactions with Earth from different perspectives

Solar Dynamics Observatory — sdo.gsfc.nasa.gov — Satellite observations of the Sun

Photo Credits